Public Speaking for Engineers

Other Titles of Interest

Becoming Leaders: A Practical Handbook for Women in Engineering, Science, and Technology, **by F. Mary Williams and Carolyn J. Emerson** (ASCE Press, 2008). Summarizes the best research on a wide range of topics of interest to women in different stages of their careers, presenting timely information along with practical tips. (ISBN 978-0-7844-0920-6)

Economics and Finance for Engineers and Planners: Managing Infrastructure and Natural Resources, **by Neil S. Grigg, Ph.D., P.E.** (ASCE Press, 2009). Presents the core issues of economics and finance that relate directly to the work of civil engineers, construction managers, and public works and utility officials. (ISBN 978-0-7844-0974-9)

Engineer to Entrepreneur: Success Strategies to Manage Your Career and Start Your Own Firm, **by Rick De La Guardia** (ASCE Press, 2016). Provides the aspiring entrepreneur with practical steps and guidance at key career points to advance your career and reach your professional goals in any engineering discipline. (ISBN 978-0-7844-1441-5)

Engineering Ethics: Real World Case Studies, **by Steven K. Starrett, Ph.D., P.E.; Carlos Bertha, Ph.D.; and Amy L. Lara, Ph.D.** (ASCE Press, 2017). Offers in-depth analysis of real-world engineering ethics cases studies with extended discussions and study questions. (ISBN 978-0-7844-1467-5)

The 21st-Century Engineer: A Proposal for Engineering Education Reform, **by Patricia D. Galloway, Ph.D., P.E.** (ASCE Press, 2008). Sets out nontechnical areas in which engineers must become proficient and issues a clarion call to reform the way today's engineers prepare for tomorrow. (ISBN 978-0-7844-0936-7)

Public Speaking for Engineers

Communicating Effectively with Clients, the Public, and Local Government

By Christopher A. "Shoots" Veis, P.E.

Foreword by
Mark W. Woodson, P.E., L.S., D.WRE

Library of Congress Cataloging-in-Publication Data

Names: Veis, Christopher A., author.
Title: Public speaking for engineers : communicating effectively with clients, the public, and local government / by Christopher A. "Shoots" Veis, P.E. ; foreword by Mark W. Woodson, P.E., L.S., D.WRE.
Description: Reston, Virginia : ASCE Press, [2017] | Includes index.
Identifiers: LCCN 2017005083 | ISBN 9780784414729 (soft cover : alk. paper) | ISBN 9780784480526 (PDF) | ISBN 9780784480533 (ePUB)
Subjects: LCSH: Communication in engineering. | Communication of technical information. | Public speaking.
Classification: LCC TA158.5 .V45 2017 | DDC 808.5/102462–dc23 LC record available at https://lccn.loc.gov/2017005083

Published by American Society of Civil Engineers
1801 Alexander Bell Drive
Reston, Virginia 20191-4382
www.asce.org/bookstore | ascelibrary.org

Errata: Errata, if any, can be found at https://doi.org/10.1061/9780784414729.

ISBN 978-0-7844-1472-9 (print)
ISBN 978-0-7844-8052-6 (PDF)
ISBN 978-0-7844-8053-3 (ePUB)
Manufactured in the United States of America.

24 23 22 21 20 19 18 17 1 2 3 4 5

Contents

Foreword by Mark W. Woodson, P.E., L.S., D.WRE vii
Preface ix

Chapter 1. Introduction 1
All Too Common 1
More Art Than Science 5
Through the Lens of Local Government 7
Through the Lens of Civil Engineering 7
Case Study: Arthur's Story 8

Chapter 2. Planning 11
Defining the Scope of the Presentation 12
Choosing an Objective 13
Selecting the Type of Speech 17
Case Study: A Tale of Two Presentations 19
Identifying the Audience 21
Case Study: Complete Streets 34
Assessing the Setting 36
Case Study: Murphy's Water Tank 45
Types of Meetings 46

Chapter 3. Design 51
Moving from Planning to Design 52
Developing an Outline 53
Case Study: TEDx Speech 59
From Outline to Speech 60
Case Study: Three-Minute Speeches and Their Outlines 69
The Speaker's Toolbox—Visual Aids 78

Case Study: A Slow, Painful Death by PowerPoint, aka the Water and Wastewater Master Plan Update 87

Chapter 4. Delivery 91
Getting from Design to Delivery 92
Assessing Your Speaking Skills 94
Skills, Strategies, and Habits 99
Speaking Skills 100
Case Study: Mind over Butterflies 113
Speaking Strategies 114
Case Study: Technical Proficiency versus Communication Skill 122
Speaking Habits 123
Case Study: A Missed Opportunity 130

Chapter 5. Local Government 133
Getting to Know Your Local Government 133
Government Hierarchy 134
Case Study: A Day in the Life of a Local Elected Official 138
Funding Mechanisms 140
Learning Local Government: Who and How 143
Case Study: The Engineer versus the Neighbors, or Subdivision Access versus Sewer Alignment 145
Engineers as Advocates for Infrastructure 147
Case Study: Over the River and through the Woods to the Municipal Wastewater Treatment Plant 150
Epilogue 152
Appendix: Resources to Understand Local Government 153

Index 155
About the Author 159

Foreword

This is a book that every engineer should read and keep as a ready reference. I wish that I'd had this much "how to" when I was starting out in practice.

No, we don't all like to speak in public. And no, we don't all have to do it in the course of our careers. But what if you could get a little better? A little more confident? This book will help you out—not only in your career, but also in many things you do within your community.

In his book, Shoots Veis walks us through the steps to be a better public speaker. And he writes it for civil engineers—though it can be applied to anyone. By writing it for engineers like us, he puts us in contexts that can help us get better at what we do, or are afraid to do.

The book lays out the fact that public speaking is more of an art than a science, but goes on to explain how we can all get better. The book talks about the various reasons for giving a public presentation. It then goes further and lays out great planning tools. And once the plan is in place, it takes you through the steps to design not just the speech, but the entire presentation.

The book then gets into the delivery and the many elements that will help you make a better presentation: from speaking skills to strategies.

It talks about knowing your audience and how to prepare for different types of audiences. It explains how to differentiate between speeches to other engineers versus talking to a local neighborhood group or government body; these are important lessons to learn.

The book goes so far as to provide an overview of how local governments are organized, which is a great tool for those who haven't already worked in this area.

Throughout the book, Shoots provides case studies to further explain the lessons that he's laid out in the book. These help to get a real feel for speaking situations and how to handle them.

As a civil engineer, a former city engineer, a consultant who makes regular presentations to city and county governing bodies and staffs, and a former city council member who listened to all manner of presentations from engineers and other professionals, I can tell you that Shoots has done a great job of laying out different scenarios that you'll face, and he's taken the time to walk you through the steps necessary to make a great presentation.

I consider myself a pretty good public speaker, but reading this book has helped me understand how I can do a better job and, I hope, get better results with my talks.

If you're at all interested in becoming a better public speaker, get this book, read it, and use it!

Mark W. Woodson, P.E., L.S., ENV SP, F.ASCE
2016 ASCE President
Woodson Engineering and Surveying, Inc.
Flagstaff, AZ

Preface

Several years ago, a friend asked if I would put together a presentation for an engineering conference. She wanted me to provide a session on public speaking. She asked me to speak because, at the time, I was serving on the Billings, Montana, city council as well as working as an engineer. Little did she know that public speaking by engineers was a topic that was piquing my interest. In my role as a city council member, I repeatedly saw poor presentations by engineers and did not enjoy the consequences of those bad speeches. So I said yes, prepared for the presentation, and delivered a well-received speech at the conference.

After the presentation, the topic of public speaking for engineers would not leave me alone, and eventually I decided to write a book about it. I believe it is important for engineers to understand why public speaking is an essential skill and how to improve their ability to do it. Chances are, if you made the effort to pick up this book, you have reached the same conclusion that I did: you are an engineer and you realize that you need to be a more proficient public speaker. Your reasons could be any or all of the following.

It makes your job easier. The better you communicate with an audience, the easier it will be to get a project completed in an efficient and effective manner. In too many city council meetings, I wanted an engineer to provide information in a way my fellow council members (who were not engineers) could understand. More often than not, I ended up translating engineering jargon into city council jargon. What happens when there is no engineer on the policy body? Engineers need to speak effectively or risk complicating their projects for lack of good communication.

It makes you look good. When you give a polished and effective presentation, it instills confidence in the audience that you know what you are talking about. Most audiences lend credence to the testimony of an engineer concerning the technical details of a project. Unless they have reason to distrust you, the audience will likely default to a position of taking your advice seriously. However, when you give a poor presentation, it shapes the perception that you don't know your topic. Although most people believe that engineers know their stuff, doing a poor job of relating pertinent knowledge shakes that confidence.

It lends an engineering voice and sound engineering judgment to policy decisions. Good public policy and decision making often *need* to be influenced by engineers. If engineers communicate ineffectively with local government officials, an important part of the discussion is overlooked, which leads to less-than-ideal public policy decisions. The better you are at speaking to policy bodies, the better the decisions those bodies will make. It is important that you are able to lend your logical, well-thought-out analysis to policy discussions. No one in local government or the engineering profession wants to see poor public policy. Engineers—and I mean *you*—must be a part of that discussion if all of us are to benefit from our local governments making good decisions about our future.

It is good for engineering jobs. If you succeed at convincing policy-makers and the public that investment in infrastructure makes sense, you sustain and help create jobs in the engineering industry.

It raises the prestige of the engineering profession. When you make a good presentation, the audience is more likely to think highly of engineers in general and be receptive to the next presentation on an engineering subject. Developing a reputation as a good presenter sure beats talking to an audience that is dreading your speech.

An engineer who can articulate a clear message shows pride in the profession. D. Wayne Klotz, 2009 President of ASCE, put it this way in his inaugural address:

> Here's a story. The story is about a man who came upon a group of masons working beside a road. He asked the first man what he was doing. "Stacking rocks," was the immediate reply. Finding that response inadequate, he inquired of the second man. "I am building a wall," was the second answer. Again, our traveler was not fully informed of the nature of the work he was observing. Asking the third mason, he received the definitive response. The third worker proudly proclaimed that he was "building a cathedral." What a contrast. Three masons working together with totally different ideas of what they were doing. Which one was right? Which one understood the value of his work?

> Ask most practicing civil engineers what they do for a living. You will hear all sorts of responses. "I design water lines." "I review subdivision plats." "I prepare environmental permit applications." My favorite answer, which I really heard about 20 years ago, was "I am into sewers." Most of these answers fit into the stacking rocks or building walls category. Is it any wonder that we must work so hard to convey the critical need to increase funding for civil works to our elected officials? Many of us do not seem to believe our profession matters, that it is critical to public health and safety, to economic strength, to our way of life. We do not seem to believe it ourselves. Well, we better believe it!

If engineers are going to change the perception from "technical expert—poor speaker" to "good communicator and trusted ally," then all engineers must do their part to become better public speakers. Each time an engineer speaks in front of a local government body or other audience, he or she gives an impression of the profession as a whole. When enough engineers give poor presentations, a stereotype emerges of a profession lacking communication skills. On the other hand, if engineers become skillful speakers, people will accept engineers not only as proficient communicators, but also as trustworthy authorities with important, relevant information.

About This Book

So you are an engineer (it doesn't really matter what kind), and you want to be a better public speaker. I wrote this book to help you do that. Recognizing that public speaking may not come naturally to us engineers, I tried to make this book easy and fun to read. I provide tips and tricks for good public speaking, but also a lot of examples and case studies. (You might even see yourself or your colleagues in some of them.)

For example, throughout this book, you will meet two characters—Tom and Vivian—whom I developed to show the principles in practice. Tom is an engineering manager, and Vivian is an engineer early in her career. Through Tom and Vivian, I can present real-life scenarios that involve engineering and public speaking. They bridge the gap between straightforward advice on public speaking and a demonstration of how public speaking becomes a part of an engineering job. Tom and Vivian illustrate how the "rules" for public speaking can actually play out in a career.

You also will find several case studies in this book. The case studies are drawn from my experience as an engineer and local elected official, and they demonstrate both what to do and what not to do. Unlike Tom and Vivian, whom I made up, the case studies come from real-life events that I witnessed. They are cautionary tales culled from my experience as an engineer and a city official. If you are wondering if they really happened, I can assure you that they did.

Acknowledgments

My thanks to all who helped me complete the book. A special thank you to my wife for helping with the editing and for all of her encouragement. Many thanks to my parents and brother for their help in editing the book. Thank you to Christy Foster for developing the illustration on the Road Cross Section slide. Thank you to Betsy Kulamer at ASCE Press for the help and patience with the process. Thank you to TL and Nancy Hines for their insight into the book publishing process.

Chapter 1

Introduction

I don't think much of a man who is not wiser today than he was yesterday.

—*Abraham Lincoln*

All Too Common

The room was a bit chilly and the air was fresh when the city council meeting began. After three and a half hours, though, with a packed crowd of people waiting for a turn to speak, the atmosphere is stuffy, warm, and stale. The council meeting has already dealt with two contentious hearings, the first a proposed ordinance to limit dogs in city parks, the second about a commercial development that will have a big impact on a well-kept, long-established neighborhood. Both hearings had several people testify on either side of the issue and close votes—banning dogs that weigh more than 85 pounds from city parks and allowing the developer to move forward with the commercial development.

After a ten-minute break, it is time for the third big hearing of the evening, to determine whether the city will move forward a proposal to upgrade a marginally congested, but getting busier by the day, arterial road. The public works director opens with a brief project overview. He invites the project engineer to the podium to give a presentation on the proposed road upgrade. Dread sinks over most of the council members: they have heard from this engineer before, and they know they are in for a long, complicated, confusing report updating them on the progress of the road. Three minutes into the engineer's speech, most of the council members tune out and stop paying much attention to what the engineer is telling them.

Unfortunately, this reaction to an engineer's presentation is all too common. Having served on the city council for a five-year term, I have attended plenty of meetings, hearings, and technical presentations, and I have seen firsthand the foggy results that follow. Being an engineer myself, I have sympathy for engineers required to speak at local government meetings. Most people—not just engineers—struggle with public speaking. And engineers have a particular burden because often their subjects *are* technical, complicated, and maybe even about things (like sewage or rights-of-way or soil conditions) that other people do not want to think about. Yet the expertise and perspective that engineers bring to a variety of problems is critical if local government groups (made up mostly of nonengineers) are to make sound decisions that protect public safety and improve public welfare. So to create better results for engineers and their audiences, I wrote this book to help engineers do a better job of preparing and delivering effective presentations. But before I get into the details, let's see how the hearing on the proposed road upgrade turns out.

After droning on for just over 15 minutes, the engineer ends his presentation. He then asks the council for input on several issues, including the type of road cross section they prefer, exactly where the project terminus should be, whether or not a trail crossing should be added to the road, and depending on the results of those decisions, how much money the council is willing to spend on the project. The engineer is asking the council for input on these questions, but most importantly, he hopes to get buy-in from them. With council buy-in, he can proceed with potentially controversial plans knowing the council will back those plans once they are finalized. Unfortunately, the engineer is oblivious to the fact that none of the council members paid attention to his remarks.

When the council questions the engineer, they ask about the parts of the project that matter to them and their constituents. Important project facets are overlooked because no one focuses on the complex aspects. Instead, small, tangential issues consume a large portion of the question session. One council member asks where construction is proposed to begin and end. The engineer repeats his recommendation, but notes that the beginning and end points of construction are still flexible. He says that council input is important to decide the limits of construction. However, no council members follow up to give advice. The engineer walks away from the podium with no clear direction from the council as to where construction should begin and end—and no direction on several other key decision points.

Next, the general public has a chance to speak about the road project. The comments go like this:

- The leader of a neighborhood group testifies that the project has come up a couple of times at neighborhood meetings and the group would like to see some changes in how the road is being planned.

They want some alignment changes and the addition of a noise barrier on the far end of the project. It is clear from his testimony that the only source of information he has on the road are the stories that have been in the local paper.
- A lawyer representing a commercial interest near the road requests that additional access be granted from the road to the business and that the road alignment be moved because her client does not want to give up as much right-of-way as is being proposed.
- Advocates for fewer commercial signs in the area speak next, noting that the road would look much better when completed if no commercial signs were allowed within 200 feet of the road and maximum sign height were 15 feet. They request that the council add this condition to the construction of the road. City staff take the time to interrupt at this point and note that commercial signs are regulated by the city sign and zoning code. They state that the conditions of road construction are not the appropriate arena to deal with commercial sign restrictions.
- Finally, an older gentleman gets up to testify. He supports the construction of the road, but as a longtime area resident, he would like to see green boulevards as part of the road design.

Overlooked in all of the public testimony is guidance for the engineer on how to proceed with the road. None of his questions, save the one about the points of beginning and ending the road construction, have been mentioned during the public testimony portion of the hearing.

The council now calls on the engineer to reply to the testimony given during the public hearing. Several council members pepper the engineer with questions pertaining to the requests of the people who testified during the public hearing. The engineer does a poor job of explaining how the extra work proposed in the public hearing could be incorporated into the project and how much it might add to the road construction cost. Rather than answering questions directly, he refers to different sections of a large engineering report that justifies the preliminary road design. To the council, it seems like the engineer is avoiding giving answers to what they see as fairly basic questions.

Without understanding the engineer's explanations as to how a delay may affect the project and without tackling any of the issues the engineer needed to have resolved, the council votes to delay any decisions on the road for 30 days. During that time, they want the engineer to host another public meeting that is well advertised. They also want the engineer to come to a less formal city council meeting and go through the project with them again. They give some general suggestions as to what they want to have covered in the additional meeting.

As the mayor brings the hearing to a close, all stakeholders—the engineer, the council, and the public—come away frustrated: the engineer, because the 30-day delay most likely means the loss of a construction season and increased construction cost; the council, because they have to spend more time deliberating about the road; and the public, because they do not understand why their suggestions have not been incorporated into the road design.

As a city council member, I have sat through meetings like this plenty of times. I have watched a straightforward project run off the rails and head for a delay. The delay may not be a bad thing for the public or the council because it gives them a chance to become educated about the project and make good decisions about how to proceed. Often, this is how government progresses, slowly and deliberately.

For the engineer, though, the delay and deliberation can put a real crimp in project plans. In this instance, the engineer had a schedule for the project, which most likely did not include a 30-day delay in getting approval from the city council. Instead of finalizing bid documents, the engineer will spend the next month preparing for and conducting meetings with the council and the public. Most likely, he also will be redesigning several aspects of the project as he gets input from the council and the public.

Admittedly, this engineer's problems cannot be traced solely to a poor presentation in front of city council. He also wasn't sufficiently attentive to the needs of groups interested in the road construction. However, a good public presentation might have ameliorated some of the concerns raised at the meeting and given the city council confidence in their engineer. The engineer would have educated the council and public stakeholders *and* walked away with authorization that would keep the project on schedule.

The purpose of this book is to give you, the engineer, insight and skills for building and delivering effective public presentations. It will help you avoid the scenario that played out during the council meeting described here. A public speaker who spends time preparing a good speech and delivering a message tailored to the audience can succeed in helping listeners understand and make informed decisions about complex, technical issues.

More Art Than Science

According to the Wikipedia entry on *glossophobia* (better known as speech anxiety), surveys have shown most people fear death less than they fear public speaking. About 75% of the population suffers from some sort of anxiety when speaking in public. In this, engineers are no different from the rest of the population. If it's any comfort, I can assure you that it is possible to alleviate some of the anxiety that goes with speaking to an audience and to get better at public speaking. Before I explore the nuts and bolts of how to do this, I want to explain an important factor about becoming a better speaker. Unlike engineering, public speaking is more art than science.

A good engineer who decides to tackle the problem of becoming a proficient public speaker may start by searching for a formula. However, high-quality public speaking is not so much grounded in the hard and fast rules of science as it is governed by the looser standards of art. With a little study, anyone can understand that force is equal to mass times acceleration, but there is no single formula that produces a great public speaker. Rather, like the standard of good art, you know it when you experience it.

You may believe that there is little you can do to become a better speaker, but that is not exactly the truth. There is no decree that you are born either a good public speaker or a poor one. If you practice public speaking skills, you will become a better public speaker, just as someone practicing painting will become a better painter. The trick for engineers is that the parameters for public speaking are looser than the technical standards to which they are accustomed. If you are looking for precise directions that will make public speaking a breeze, well, that manual doesn't exist. If you want to become a better speaker, you can go a long way toward success by working hard to prepare your speech and then working equally hard to develop the skills and habits associated with effective public speaking.

The "art" of public speaking likely pushes you out of your comfort zone. Chances are, you did not become an engineer because of your desire for public appearances, flair for the dramatic, empathetic sense of an audience, or ability to sell anything to anyone at any time. I expect, instead, that you like to take solid principles and apply them logically to a problem.

Me too. Those are good traits for engineers. We want our engineers to build bridges that do not collapse, water systems that deliver drinkable water, airports that support safe travel, and electrical systems that provide reliable electricity. Engineers rely on logical judgment and proven experience to make decisions. Logical problem solving is taught to engineers from the moment they step onto a college campus until the day they leave with mortarboards on their heads. We want our engineers to be detail-oriented problem solvers. Engineers would be great public speakers if there were a mathematical formula to follow. Sadly, public speaking does not work this way.

I'm not saying that all engineers are poor public speakers. I've heard excellent ones who do not need any coaching on public speaking. I've listened to adequate ones who could muddle through a speech with mixed results. However, far more often, I've heard engineers who are poor public speakers and could use help improving their abilities.

Luckily, public speaking is a skill that you can work on and improve. As you work to develop your speaking skills, just be aware there is no scientifically based, logical, technical explanation of exactly what you are supposed to say in every situation. It would be easier to find Atlantis. However, following some basic rules will go a long way toward improving your abilities. In this book, I cover the following principles of good public speaking.

Planning

- Objectives,
- Audience,
- Setting, and
- Meetings.

Design

- Developing a good outline,
- From outline to speech, and
- Effectively using presentation tools and visual aids.

Delivery

- Assessing your speaking skills, and
- Improving your public speaking skills and habits.

As you begin working on becoming a better public speaker, remember that you will be out of your comfort zone, that there are no hard and fast engineering principles to follow, and that public speaking skills are more art than science.

Through the Lens of Local Government

All of us are influenced by the forces and circumstances around us. If we grew up in a place with a strong sports tradition, we tend to follow the team that makes its home in that area. If our parents practiced a certain occupation, we may follow them into that occupation. My life has been influenced by spending a fair amount of time around government bodies and a term as a city council member. On the council, I saw engineers and their presentations the way nonengineers saw and heard them. I began to understand the ramifications when an engineer can't communicate effectively to audiences who aren't also engineers.

It can be a daunting proposition to stand in front of a group of elected officials, each with a nameplate identifying them and their title, and speak about something you know a lot about and (maybe) they don't. Throughout the book, I will provide insight into the audience members looking at you from the other side of the nameplates. I will explore how to prepare a presentation, how to understand an audience and address its needs, and how to deliver an effective presentation. I hope to enable you to succeed when you speak in front of local government bodies or other audiences. Keep in mind that my perspective comes from my time spent as a local elected official. That experience is an important lens through which I view this topic.

If you are not often called on to speak to the audiences or in the settings I describe, you will still be able to draw some parallels between the scenarios dealing with local government and the settings in which you speak. Sometimes this work may be easy; sometimes it may be a struggle. Good preparation, audience understanding, and effective speaking skills are universal themes of public speaking, however, and they are applicable any time a speaker stands up in front of an audience.

Through the Lens of Civil Engineering

Another significant influence in my life is my profession as a civil engineer. Most of the stories I share are colored by the type of engineering practiced by civil, traffic and road, and environmental engineers. After all, these engineering disciplines work most closely with local governments and are often called on to speak at public meetings. That said, the structure and preparation I describe for public speaking can work for anyone who struggles with public speaking. Just be aware that the profession of civil engineering is a lens through which the information is being passed.

Case Study: Arthur's Story

Arthur, a midlevel manager in a decent-sized city, was good at his job, managed his people well, and enjoyed the work. Although he was within five years of retirement, he quietly let it be known that he would like to be department director. Because he was well liked, it was not a stretch to believe that he could fill the position for a couple of years before he retired. He was motivated by the prestige of running the department he had worked in for 25 years and the added bump to his retirement salary.

When the existing department head took a job in another community, Arthur was promoted to interim director until a new department head could be selected. Being a government job, the position could not just be given to Arthur; rather, a competitive hiring process had to be followed. Arthur gladly took the interim director position during the four-month hiring process and applied for the permanent job.

During the hiring process, it became clear that Arthur would not be chosen as the permanent department head. Although he was competent at running the day-to-day department operations, he was a disastrous public speaker. As a midlevel manager, he had not been required to do a lot of public speaking, nor was he comfortable speaking to groups. Unfortunately for Arthur, the public speaking skills he lacked were required to be the department head.

Council members who heard him give testimony at a council meeting knew immediately that Arthur would not get the promotion he wanted. At a public meeting, getting a simple answer out of him was like pulling teeth. One on one, he was better at providing answers, but put him in front of an audience and he would quietly babble on and on. Council members and other department heads felt sorry for Arthur when he was called upon to speak at a public meeting, but they could not wait for a new department head to be hired, one who could address them in a meaningful way.

Arthur himself struggled in public meetings. Audience members asked him the same questions over and over because he was not able to articulate a clear answer. He avoided opportunities to speak to the public about projects outside council meetings, with the result that members of the public showed up claiming they had never been given the details about projects of which Arthur was in charge. This caused problems for the council, which then had to make changes to meet the expectations of the ill-informed neighbors.

In the end, hampered by his inability to do an adequate job of public speaking, Arthur didn't get the promotion he wanted. In a couple of years, Arthur retired, having never achieved his goal of becoming department head. Most definitely, he had been held back by his lack of skill at public speaking.

Chapter 2

Planning

There are always three speeches for every one you actually gave. The one you practiced, the one you gave, and the one you wish you gave.
—Dale Carnegie

E-mails arrive in an engineer's in-box throughout the day, every day. Most concern projects or office information. Every once in a while, an e-mail arrives that tells the receiver that she needs to prepare a public presentation. This chapter starts with just such a scenario playing out between two fictional engineers, Tom and Vivian. Their scene sets the stage for the information about preparing a public presentation that makes up this chapter. The discussion then turns to each of the factors that should be analyzed as part of an effective speech plan:

- Define the objectives for your speech so that you can focus on what you are going to say to your audience.
- Determine what type of speech you will be giving. Most speeches fall into three categories, but the one often used by engineers is a speech to inform.
- Identify who will be sitting in your audience so you can gauge how best to communicate with them. Audiences come with many different characteristics, and an overview of some different audience types is provided.
- Assess the room where you will be making your presentation. Your preparation should include investigating how best to communicate in that room. You need to know how the room will be arranged and what audiovisual equipment will be available for your use. It is also important to plan for problems and develop a Plan B in case something goes wrong.
- Classify the type of meeting you will be addressing. Like the variable audience characteristics, different types of meetings require different presentation types. Formality that is appropriate in one situation will seem stuffy in another.

This chapter includes three case studies. The first case study is on presentation planning and how it affects delivery. It describes a situation where two presentations were given back-to-back, one done well and the other done poorly. The second case study describes a meeting that did not go well because the speaker had not prepared to reach his audience. Granting that it was a tough crowd, the speaker had not considered the audience objectives when he prepared his presentation. Thus, he was unable to steer the audience past their differences toward a solution that satisfied all parties. The third case study describes a presentation with a potential client that did not go well because no effort was made to develop a back-up plan in case things went wrong. Several elements of the presentation were disrupted, but could have easily been solved with a little planning.

Defining the Scope of the Presentation

That tinny bell noise signifying an incoming e-mail goes off again, and Vivian glances up to see what the cyber postman has delivered. The e-mail is from the boss, with a subject line that says: "City Council presentation." Not all that excitedly, Vivian decides she had better open the message to find out what is in store for her.

> Vivian,
>
> I just spoke with the Public Works director, and now that we have selected the preferred alternative, he would like you to give a presentation on the Main Street road reconstruction project to the city council on Tuesday, Feb. 3rd. Stop by my office later today and we'll go over what you will need to cover.
>
> Thanks
>
> Tom

Not many engineers would consider this e-mail good news. Most of us would be hesitant about giving a speech on a project. It is part of an engineer's job, however, and an important aspect of any project. Effectively communicating the scope, budget, time line, and other aspects of a project to an interested public or a group of local government officials goes a long way toward ensuring success. After stopping by her boss's office to get the details, Vivian considers what she needs to do to prepare a good presentation.

If you find yourself in Vivian's shoes, the best place to start is by focusing on the presentation objectives. This process is similar to developing a scope of work for an engineering project. A project scope of work involves outlining the broad strokes of what the project should accomplish and defining the project goals. A scope of work is developed at the beginning of a project to put the project on the correct path, so that appropriate decisions are made at each step along the way. A speaker develops the objectives and outlines the speech to put them in the correct order. The purpose of a road project is vastly different from that of an electrical transmission line or a shipping port upgrade. The planning process at the beginning settles the project objectives and provides an outline of what the project will accomplish.

In order to impress Tom and the city council with her presentation and communication skills, Vivian will establish her objectives, develop an outline, and determine what type of speech will foster communication with the audience. She will assess how to reach her audience by understanding their needs and how to successfully speak within the confines of the room. Last but not least, she has to communicate her objectives in the assigned time span.

Choosing an Objective

Effective communication with the audience begins when you, the speaker, have a clear vision of what you want your listeners to hear and understand. An ambiguous message from an unprepared speaker ensures speech failure.

As you are starting to think about your presentation, make notes as to what you wish to communicate. Begin by listing the one objective you want to accomplish. If there is more than one, then put the additional objectives on the list. Later on, you will match the number of objectives to what the audience wants to hear in the amount of time available for the speech. But figuring out your aims is a crucial first step. Bear in mind that a good speech most often has no more than three objectives.

Next, list the information that needs to be passed along to the audience. Spend some time thinking about what aspects of the project are relevant to the audience and what they want to hear about. If you have trouble determining the presentation objectives, ask the person requesting the speech what she believes the audience wants to learn. This should guide you as to what topics you should cover. It is important to match your aims for the speech with the audience's goals in listening to it.

Be aware that, initially, you may identify five critical objectives that must be covered in the speech. Later on, you might find out the speech will be a little over 10 minutes long and the audience is really interested in only two of the objectives. At that point, you will have to adjust your speech. However, you have to begin somewhere, and knowing what you want to accomplish is a great place to start.

Once you have the list of objectives, you will need to compare them with the amount of time allotted to discuss them. If you are speaking at a conference and are given an hour time slot, you can cover several key points. When you are limited to about five minutes, you are going to have to whittle your list to one or two key points. If you are unsure of the expected speech length, then ask.

Once you have determined what to say and how long you have to say it, give some thought about how to best relay the message to the expected audience. An effective public speaker finds a way to communicate key points so that the audience will remember them later, long after the speech is over.

If you get a request similar to the one Vivian received at the beginning of this chapter, decide what your objectives are. Have a few in mind before heading down to your manager's office to discuss the presentation. Visit with the manager to test your objectives and ask for insight about the audience and its expectations. The Planning Checklist (Fig. 2-1) can help you organize what you learn in this chapter and provides a space to write down the objectives of the speech.

* * *

Vivian begins her planning for the council presentation by writing down the objectives she wants to include in the speech. She wants to be thorough and cover all aspects of the project. Her initial list includes the following:

1. Deliver preliminary construction cost estimate.
2. Discuss funding sources.
3. Discuss construction time frame.
4. Discuss why the preferred alternative was selected.

Planning Checklist

Fill in the blanks and circle the options that apply

Three Objectives of the Speech:

1. ____________________

2. ____________________

3. ____________________

Type of Speech	Inform	Persuade	Entertain
Audience	Colleagues	Students	Elected Officials
	Government Staff	The Public	Board Members
	Conference Attendees	People I Know	Selection Committee
	People I Don't Know	Engineers	Acquaintances

The Audience WILL / WILL NOT be able to write during the presentation

Speech Location

Physical Address: ____________________

Room Number: ____________________

Maps:	Road Map	Conference Center Map	
	Building Map	Parking Map	
Type of Room:	Conference	Classroom	
	Board Room	Local Government Meeting Room	
The Room has:	Desks	Tables facing the speaker	
	Round Tables	Rows of chairs	
	Good Acoustics	Bad Acoustics	
The Room is:	Large	Small	
	Warm	Cold	
The Room has:	A Podium	A Microphone	A Computer
	A Projector	Plug-Ins	Internet Connectivity
Plan B	Plan B Checklist IS / IS NOT complete.		
Meeting Type	Local Government	Kick-off	Conference Presentation
	Committee	Interview Committee	Community
	Staff	Sales	

Other: ____________________

____

Fig. 2-1. Planning Checklist

5. Graphically show the beginning and ending of the project and what streets will be affected.
6. Cover the next steps for the project and decision points.
7. Highlight which businesses will be affected by road construction.
8. Graphically show how the detours will be set up and where the barricades will be located during construction.

Having come up with eight objectives for her speech, she will need to narrow the list of objectives to three to fit in the five-minute time frame. Looking over the list, she realizes that objectives 1, 2, and 3 can be combined into one objective with one slide. A lot of technical engineering analysis was used to determine the preferred alternative (objective 4), and Vivian recognizes the council may not be interested in hearing the detailed information on why the preferred alternative was chosen. She decides to eliminate objective 4.

Vivian does want to provide a graphic during her presentation that shows the project's geographic beginning and end points. Looking over her objectives, she realizes that the same graphic can also be used to show which businesses will be affected by the project. She decides to eliminate objective 7 and incorporate the businesses affected into the graphic she uses for objective 5.

Upon reflection, objective 8 does not seem like something that will interest council members. Vivian hopes they will trust her engineering skill to set up the detours and will not have questions about them. She chooses to eliminate objective 8. Finally, she decides that objective 6 should probably be the last thing she talks about, so she places it at the end of the list. Vivian's edited objective list looks like this:

1. Deliver preliminary construction cost estimate, discuss funding and time frame.

~~2. Discuss funding sources~~

~~3. Discuss construction time frame~~

~~4. Discuss why the preferred alternative was selected~~

5. Graphically show the beginning and ending of the project and what streets and businesses will be affected.
6. Cover the next steps for the project and decision points (move to last objective).

~~7. Highlight which businesses will be affected by road construction~~

~~8. Graphically show how the detours will be set up and where the barricades will be located during construction~~

She cleans up the list to illustrate a clear sense of the speech objectives. She can use this list to begin developing her presentation.

1. Deliver preliminary construction cost estimate and discuss funding and time frame.
2. Graphically show the beginning and ending of the project and what streets and businesses will be affected.
3. Cover the next steps for the project and decision points.

Selecting the Type of Speech

With a good understanding of your speaking goals, you next identify the type of speech you will be giving. There are three general types of speeches:

- Speeches that inform,
- Speeches that persuade, and
- Speeches that entertain.

Engineering presentations most often fall into the category of speeches that inform. However, engineers also are called upon to give presentations to persuade audience members to choose one option over another. Admittedly, few engineering presentations are given for entertainment value; they simply do not lend themselves to drama or comedy.

A Presentation to Inform

During a presentation to inform, a knowledgeable speaker communicates facts or technical expertise on a topic to an audience that needs the knowledge (whether they know it or not!). A presentation to inform should be pretty straightforward. However, I caution against using a Joe Friday, just-the-facts-ma'am approach accompanied by reading a collection of slides to the audience. A good, memorable, informative speech is never a recitation of every detail on a given topic. Rather, it is a well-structured presentation containing material that the audience can relate to and learn from.

Your presentation should provide information proficiently, graciously, and in a way that interests the listener. It should be clear, concise, and connected to the audience. Focus on providing details that the audience does not already have. If they have some basic knowledge of the topic, then provide them with new insights or perspectives. Organize the presentation so that the objectives can be conveyed with ease and retained by the audience. Try to avoid complex technical terms and engineering jargon. Make sure your audience will clearly understand and remember your objectives.

The basic structure of a speech to inform is to provide a linear progression through the information the speaker wants to provide to the audience. It should be organized like this:

Intro → Body → Conclusion

The body of the speech should contain the objectives of the speech, and when the objectives have been covered, should move on to the conclusion.

In Vivian's case, she will be giving an informational presentation when she speaks to the city council. Her goal is to educate her audience on the project's status. She is updating the audience on three aspects of the project by informing them about the cost and time line, project boundaries, and next steps. Her role as speaker is to present straightforward information to update the council on the project process.

A Presentation to Persuade

A persuasive speech, as you may have guessed, is one that attempts to convince the audience of a certain point of view. These can be challenging, especially if the audience comes into the presentation with established viewpoints. Your job as speaker is to get them to change their minds, or at least to see things from a different vantage point. A good persuasive speech will use an appeal to either ethical reasons, logical reasons, or emotional reasons for the audience to change their minds.

Changing someone's mind through a speech requires good planning, a firm grasp of the facts, knowledge of the strengths and weaknesses of your argument, and the ability to connect with your audience. Persuasive speaking is a difficult task that requires both a good argument and empathy for opposing viewpoints. Later in the chapter, a case study (Complete Streets) discusses a traffic speed reduction and road restriping project. In the case study, the engineer tried to convince the audience to accept the "best" solution to the problem.

A speech to persuade is structured differently. It should be organized like this:

Define the problem → Discuss the solutions → Call to action

The objectives of the speech may be contained in all three sections of a persuasive speech. One objective would be to convince the audience there is a problem, the next objective is to communicate the conclusion, and the final objective would be to ask for their assistance.

A Presentation to Entertain

If you do find yourself in a situation requiring an entertaining speech, use the same structure and planning as other types of speeches. But be mindful of your overarching mission and make sure that any humor used is appropriate for the situation. Nothing will depress an audience looking for amusement more than humor that misses its mark. A speech to entertain should be structured to meet the situation. The best entertainment speeches are stories appropriate for the occasion, and there are many different ways you can structure a story.

Each of these types of speeches is structured differently because each type has a different purpose. Each of the three speeches has objectives, but the objectives need to be placed in the speech structure.

Case Study: A Tale of Two Presentations

Any day I get to deliver a speech is a good day, so I sent in an application to speak at a conference of water and wastewater operators on a topic I enjoy presenting. My submission was for a presentation I had spent two months developing and would be able to deliver a couple of times between the acceptance at the conference and the day I would speak. My proposal was accepted, and I celebrate each time I convey the message because I want my audience to walk away with a heightened appreciation for infrastructure.

Developing the presentation was not a burden because I believe in the topic and expended the effort necessary to develop a good speech. I focused on defining the scope and objectives to fit them into a 30-to-45-minute time frame. I found just the right story to open the speech and used graphical slides to reinforce each point. I practiced several times before my first delivery and made small adjustments to the presentation after each of the first handful of deliveries. I have given the speech many times, and it has never failed to be a hit.

About a month after my initial presentation was accepted at the conference, the organizer of the conference called to say the presenter scheduled to speak after me had canceled. She wondered if I would be available to deliver another session at the conference. I jumped at another chance to speak and suggested a couple of talks I give on a regular basis. We agreed that one of them would be great, but she still needed to get it approved for continuing education credit. Over the next week, I refreshed the presentation, but she called back to say the proposal had been rejected. The topic was not eligible for water and wastewater operators' continuing education credit. The purpose of the conference was to ensure that attendees earned their continuing education credits, so we had to move on to a different topic.

At this point, I felt obligated to deliver the extra presentation, so I proposed another topic that I knew would be eligible for credit, one that I was interested in but had never presented. It was relevant to the conference, so it was decided that I would fill the hole in the schedule with a presentation on trenchless rehabilitation methods for water and sewer pipe. I had about a month to get my new presentation ready, more than enough time if I worked diligently. While I was interested in the topic, I was not particularly jazzed about doing a presentation on it, but I had made the commitment.

Diligence was not my friend, and I kept putting off defining the scope and objectives needed to assemble the presentation. When I had about a week left, I began researching and gathering the information that I wanted to include. I was able to put together some vague objectives for the presentation and decide on the topics I wanted to cover. Then I got busy with other work tasks and did not have time to put into the presentation. In my mind, that was fine because I was not motivated to work on the presentation anyway, and it was easy to make it a low priority.

By the time things got settled down at work, I only had a day to construct the presentation. I was able to put together a workable slide deck because of my hasty research on the topic, but I was not able to spend any time practicing the speech. My planning was short and unfocused, but I kept telling myself it would be fine because I am an experienced public speaker.

The conference started the next day, and I was the opening speaker. I led off with the presentation I knew well and delivered a great presentation. I could tell it was a hit by the way the audience reacted.

They were engaged, asking questions and paying attention. I led, and they followed. Several people talked to me after the presentation on various aspects and told me they enjoyed the presentation, so I knew they were listening. Evaluation comments confirmed that the audience enjoyed the speech.

The other presentation did not go well. With a loose scope, fuzzy objectives, okay slides, and no practice, I floundered as a speaker, and the audience was not engaged. I got one question, and when I looked out over the audience, most were not paying attention. I covered the material quickly, mostly relying on the slides to provide cues for speaking because I had not covered the information previously. I read a few of the slides to the audience, something I hate doing, because I had not practiced enough to know the slides. My energy level was lower, my speaking voice was less dynamic, and my audience was less engaged. I did not get questions about the presentation afterward, and no one mentioned how much they enjoyed the speech. Thankfully, there were no comments on the evaluations about the second presentation, but I know it was less effective communication.

There was a stark contrast between the two presentations I gave that day. One I worked hard to put together with well-defined objectives and lots of practice. One I threw together at the last minute. One was enjoyable for the audience and conveyed the information. The other was painful for the speaker and audience and transferred little, if any, information. The results bear out the facts—time spent organizing and outlining your presentation matters—even if you are an experienced public speaker. Practicing your presentations will give markedly different results.

That day the same speaker gave two different presentations—one good and one bad—and it was planning that separated them.

Identifying the Audience

You have to enter their world, because they're not coming to yours.
—Steven Cerri

Just about any book, website, or class on public speaking defines understanding your audience as a fundamental rule of public speaking. This means that you, the speaker, are under an obligation to prepare for the specific audience that will be listening to your presentation.

You should organize your material to suit *that* audience and understand how to communicate with them. Speakers with little knowledge of their audiences have very little chance of communicating well. It does not get any more basic than that.

For example, we have all been to see a doctor to get treatment for some sort of ailment. I have been to doctors who did a good job of explaining what was wrong with me and to others who seemed to be speaking another language. Doctors who talk to patients in the same way that they talk to other doctors end up with some pretty confused patients. These doctors are not paying attention to their audience, and the patients walk out of the office with little understanding of what ails them or how to fix it. The communication link between the doctor and the patient is broken because the patient is not given a chance to listen and understand.

Unfortunately, I have seen too many engineers make this same mistake in addressing an audience. Instead of taking time to analyze the audience, these engineers give presentations that assume the audience has spent years in the engineering field and is fluent in engineering's technical jargon. They fail to understand their listeners and the best way to convey information. Luckily, this is an easy problem to combat with some insight into the audience and some knowledge about presentation planning.

The Speaker's Role

Before you can connect with your audience, you need to understand how you relate to your topic. For instance, were you chosen to give a presentation because you are deeply knowledgeable about the topic? Have you spent significant time learning the subject? Do you have detailed and precise intimate information on the issue? Or were you invited to speak because you know at least something about the topic and you were available to give the presentation? Maybe your technical expertise is adequate, but you are really good at distilling it down and educating an audience? In order to relate to an audience, you should spend time figuring out where your expertise on the topic lies.

Next, consider your relationship with your audience. Are you a consultant who has been asked to provide expertise to the audience? In that case, you are likely to give an informative speech providing special knowledge of the subject. Or, are you, say, a member of the local government staff who is called upon to update your colleagues about a project? Your audience will want to hear an informative talk on how things are progressing. Or maybe you are called upon to persuade an audience at a service club that a local bond proposal is a good idea. You will need to tell them why the bond was proposed and what voting for the measure provides for them. If your job relates to the speech topic, your audience will want to know that.

Always remember that communication is a two-way street. If you, as a speaker, do a good job of learning about the audience and communicating with them in an acceptable manner, you have done your part in creating a communication pathway. The audience has to do its part and be willing to listen. Part of their willingness to listen will be based on how well you put together the presentation.

Keep in mind this is not an equal partnership. You, as the speaker, have the major responsibility to communicate. If you give a presentation and it fails, then that failure is almost always on your back.

The Broad Strokes

Each time you step in front of an audience, you should have a clear and precise picture of who your audience is and what level of technical information your audience will be able to understand and retain. Other factors occasionally can come into play, but those two—the characteristics and technical knowledge of your audience—will ensure that you are communicating with your audience, not just talking at them. It does no good to broadcast a well-structured and practiced speech if the audience is not able to receive the message.

Good communication with an audience requires that you understand their objectives in attending the speech. Just as you defined your objectives for a presentation, you should consider why they have chosen to sit in front of you and listen to your speech. There has to be overlap between your objectives in speaking and their objectives in listening.

Another important aspect of preparing for an audience is understanding group morale. Judging the mood of a crowd may not be the easiest thing in the world to do, but you should make an effort to read their feelings. As you get up to speak, the audience can have all kinds of different reactions. They may be looking forward to your comments or dreading them; bored to death by previous speakers; or happy or sad over an event that has affected many of them. It is possible to anticipate the crowd's mood as you prepare your speech. For instance, if you know you are toward the end of a long meeting agenda, you can predict that your audience will be looking for brevity and lightness. If you are addressing a community disaster, your audience will be somber. If you are the first morning speaker at a large conference on a hot topic, expect them to be alert and awaiting a brilliant speech.

Audience Characteristics

Any audience awaiting a presentation will have some traits the speaker should consider during speech planning. Understanding these attributes will give insight into what your audience wants to understand and how to best communicate with them:

- Experience and knowledge the audience has with the subject. If the topic of your presentation has received a lot of attention from the audience, they will feel like they are familiar with the topic. They may have misperceptions about the subject, but they believe they know something about it. Or the audience may have very limited understanding about the subject and is attending to learn more.
- Educational background. An audience full of college graduates will be listening for different aspects of a presentation than a group of high school students.
- Audience size. Speaking to a group of seven people is different than a group of 70, which is different than a group of 700.
- Primarily male, primarily female, or mixed audience. Gender differences in communication style may be important to a speaker if the audience is weighted heavily with one gender or the other.
- Age range. Speaking to a group of high school students will require a different type of presentation than speaking to a group of senior citizens.
- Participants' occupations. A presentation to a group of accountants should be prepared a bit differently than one addressed to a group of teachers. Each profession has its own characteristics, which a good speaker will prepare for and understand.
- Income range. Speaking to a group of wealthy businesspeople about a proposed project is going to be different from speaking to a group of senior citizens who are on fixed incomes.
- Whether you know anyone in the audience. People who know you will listen differently than a group of strangers.

Each presentation comes with a different audience, so a prepared speaker will evaluate the possible traits and then tailor the speech to meet the needs of that group. In addition, each new audience may present a new cross section of people, perhaps with different objectives and technical levels. So, let's cover some of the possibilities of who might be listening to your speech and how you might best communicate with them. If you have an audience with an array of characteristics, you can still think about the specific traits and how to relate to them.

Not Engineers

One of the hardest audiences for an engineer to address may be an audience of people who are not engineers. Like other professions, engineers share a common education and, often, a common way of solving problems and thinking about the world. Among themselves, engineers use professional jargon and follow occupational norms during their daily work routines. It's easy to forget that nonengineers may not have the same understanding of what engineers do and how engineers communicate.

I have listened to several presentations that began with an engineer speaking to the audience as if it had been looking over his shoulder through all of the work the engineer had done on a project. It was as if the presenter believed the audience had put everything else on hold until the engineer came along to make this presentation. All of us get wrapped up in our own work, sure, but that's no excuse for assuming the audience has been in the trenches with you during the design work on a project.

A good way to understand this phenomenon is to compare the time an engineer works on a project to the time a city council member spends on the yearly budget. While an engineer may have invested significant portions of her life on a project over six months, the council member has spent little, if any, time working on it. Instead, the council member is working on the budgeting process. He has a lot on his plate and will not be concentrating on the project like the engineer will.

Let's do a little comparison to help understand why an audience member may not be up to speed on a project. Vivian is tasked with updating the city council on the Main Street Road project. Over the past six months, Vivian has spent her time

- Developing the scope of the project;
- Ordering the land survey and CAD work to be completed;
- Developing the initial design of the road;
- Meeting with interested parties to go over initial design and get input;
- Making changes and completing the final design;
- Submitting the project for review to the public works department; and
- Finalizing the project, bidding the project, and awarding the bid.

It has taken up a significant portion of her life over those six months.

During that same six-month period, the city council members in her audience spent the time reviewing the city budget. They received reports and read documents on

- An overview of the budget;
- Public safety—fire and police;
- Public works—streets, water, sewer, solid waste;
- Airport and transit;
- Administrative functions, city court, and planning;
- Parks and recreation and the library; and
- Public hearings, budget amendments, and approval.

It has taken up a significant portion of their lives.

During those six months, they have not been following along with Vivian. They do not know the changes and decisions that have been made on the project. If Vivian wants to communicate with this audience of nonengineers, she cannot assume that they understand the engineering process that led up to the presentation. This is true not only because they are not engineers, but also because they have been focused on different tasks over the past six months.

In practice, this means that you, the engineer about to give a presentation to the city council, must step out of your project bubble and look back at it the way a peripherally involved nonengineer would. This can be difficult because you have been so deeply immersed in it. So, think about it this way: If you had not been a part of the budget process, would you want a city council member to dive right into the budget details during a presentation about the local government budget? Or would you appreciate an overview to help you shift focus and orient you to the details and the major issues?

Because engineers speak so frequently as experts on technical topics, they must always take care to bring their audiences up to speed. Give your audience the courtesy of a project overview before getting into the weeds. For instance, if you are asking the city council for input on a particular aspect of a project, start by providing an overview of the project and then reporting on progress. This way, you are more likely to get a satisfactory quick decision and avoid having the council delay action until they get a better feel for what you propose.

Local Government Officials

Local government officials are a subset of the not-engineer audience. Because civil engineers make so many presentations to government officials, let's discuss how best to communicate with them.

Most local elected officials do not seek office for personal gain, but rather to serve their communities. Serving as a local elected official is often a thankless job with long hours and low (or no) pay; many of these people are there for altruistic reasons. No matter which way they vote on an issue, local elected officials make decisions that are unpopular with some segment of their constituency. Public service is a stressful job undertaken by people looking to do some good for their hometowns. They want to lend their voices to the decisions that shape their communities and find rewards that are not monetary.

If you accept that most elected officials are there to serve the people, you have gone a long way toward understanding this audience. And your acceptance will open a pathway to effective communication with them. For instance, if you make it a point to highlight the ways the project (or whatever you are speaking on) will benefit their community and their constituents, government officials will tend to listen. By explaining your project in terms of how it benefits the people who elected them, you are striking the right chord and you are more likely to get a fair hearing. Granted, some officials may become jaded over years of public service, but by showing them the project benefits for the community they represent, you are communicating a message that will resonate.

Admittedly, every once in a while, a bad apple is elected to local office, but this is less prevalent than commonly believed. We all know that a few bad apples spoil the barrel; in this case, the occasional incompetent, corrupt, or self-serving official tends to perpetuate the perception that too many politicians are in office only to promote their self-interest. Unfortunately, these elected leaders make for juicy news items and detract from the good work done by countless individuals who have dedicated years of their lives to serving the public in an upright manner. We expect a lot from our elected officials, and it is only when they do not meet our expectations that we take notice.

So next time you prepare to speak to local government officials, take with you an understanding of why they are in the room and communicate with them by addressing their motivations and objectives.

Boards and Commissions

Most local governments have a group of boards and commissions that provide advice to elected officials. Some of the boards are established by state or local laws or codes, some are based on regional boundaries, and some are advisory (created for only a short time). Statutory boards are often created by laws enacted at the state or local level and may even have some quasi-judicial powers to make decisions. They often deal with land-use issues and make recommendations to local government bodies. It is vital to understand the role the board plays in reference to the local government to communicate effectively with this audience.

The people who serve on boards and commissions are similar to the people who serve as local elected officials. In fact, local politicians often cut their teeth serving on a local government board, so communicating with board members should be approached in a similar fashion to communicating with a local official. The board meetings usually shadow local government protocol, so the setting and location are familiar to people who follow the local government meetings. The nice thing about speaking to a board is that its mission is often narrow and you can easily identify their objectives. The Parks and Recreation Board, for instance, is interested in how your project affects parks.

You should be aware, however, that boards often have members who joined *because of* the board's narrow focus. For instance, the planning board often contains members who work as realtors and developers, partly because these people have a business interest in the board's work, but also partly because they tend to be outgoing and community-focused. Before speaking to a board, spend some time learning the board makeup and the background of board members. You can probably find this information on the local government website, but you might have to make a phone call to the staff member responsible to the board to find out who serves on it.

Sophisticated versus Unsophisticated

Before you can decide how much overview to provide and how much technical detail to go into, you need to analyze how much prior knowledge your audience is bringing to the presentation. If your audience brings a good understanding of the subject, the presentation can be more in-depth, detailed, and complex. You may be able to assume that they will know certain technical terms, so you only need to explain them briefly or not at all.

If your audience is not bringing knowledge or familiarity with your topic to the table, then you must take that into account. Without getting bogged down in too much detail, your presentation may need to focus on giving the audience a good overview and enough background that they can absorb the details. Your speech may need to be less ambitious or less complicated, but you must never be perceived as talking down to them.

Here are a couple of examples. If you are an engineering consultant specializing in traffic engineering, you may be asked to give a presentation to the traffic engineering section in the public works department of a large city. You can trust that the audience will understand complex traffic engineering concepts and objectives. They are a sophisticated audience in this context because they are used to communicating about traffic engineering issues. They will be able to discern whether your presentation makes sense or you are underprepared and winging it.

Imagine speaking on the same topic before the city council of a small rural town. Most likely, this audience will not have a lot of expertise in traffic engineering. If you approach this audience with a bunch of traffic engineering shorthand, they will not be able to decipher your message and they will leave the presentation confused and uninformed. Bear in mind that this audience is unsophisticated only about traffic engineering; they may be very sophisticated on other topics.

Internal versus External

Internal audiences are composed of people who are members of your firm or organization; external audiences include people who are not. You can expect that people from within your organization will be more familiar with you and your subject matter. They will share common reference points, understand organizational values, and have a common vocabulary. Internal audiences can be made up of coworkers, team members, professional associates, or members of a club. When you speak to them, you may be able to use some shortcuts, trusting that they will be able to follow you. But this audience is also more likely to find holes in your presentation. You may be able to prepare less, but you still need to focus on aspects of the presentation that are not held in common.

External audiences, on the other hand, do not share inside information or common background with you. With an external audience, you will want to use words and phrases that are easily understood. This audience may need more background material to bring them up to speed on unfamiliar topics. They are less likely to find your mistakes, but you will have to work harder to ensure communication because they lack shared knowledge. Preparing for an external audience requires striking the right balance between explaining unfamiliar concepts and ensuring the presentation objectives are communicated.

For example, your firm puts you in charge of developing a proposal for the engineering fees associated with a large airport runway rehabilitation project. You outline the tasks involved with design, bidding, and construction observation and estimate the number of work-hours involved. The project will have several team members, so you present the fee proposal to a group of the firm's project engineers and field personnel to allow them to discuss the proposal. They will listen to the presentation as coworkers, with an understanding of the values, experience, and culture of your organization. The presentation will spark a discussion of any changes needed to the engineering fee based on what is best for the firm and client, and the firm's experience with the challenges and opportunities realized during similar past projects. Discussions of expected profit and risks involved will be hashed over, which are unlikely to be topics appropriate for a presentation and discussion with the client.

After the proposal has been updated and everyone in the firm is happy with it, you will present the proposal to the airport board and management. The presentation given to this audience will change from the presentation you gave to the members of your firm. The internal topics of the initial presentation are not appropriate in this setting, and the airport board and management audience objectives are different. The speaker's responsibility is to adjust the presentation for the external audience.

Interdisciplinary Teams

The great benefit of being a part of a team that incorporates people from different careers is the advantage of diverse knowledge and experience. A team that includes an engineer, an architect, a lawyer, a social worker, a banker, a construction foreman, and other assorted colleagues can lead to an outstanding project. Contributions from a varied array of people are a great way to improve the engineering and all other project aspects. However, the team's shared responsibility for the project can be a recipe for disaster without superior teamwork and communication.

As a part of an interdisciplinary team, each member—including engineers—must have a high standard for consistent, clear, and professional communication among the group members. Each team member has professional jargon, so it is important that everyone speaks clearly and directly in a succinct manner that conveys professional knowledge without confusion. If you are planning a presentation to an interdisciplinary team, take the time to assess the knowledge level of team members concerning the engineering aspects of the project. A good presentation will communicate with the members clearly, match the team's objectives, and provide information that all members can comprehend. Fielding questions during and after the presentation is a valuable way to clarify and learn.

Audiences with Shared Sensibility

When you speak to an audience of people who share the same education or professional training with you, you will find that you can communicate successfully with less preparation. Your listeners probably know the basic premises of your project, and generally they will be able to follow your presentation without much explanation. They expect a complex level of communication, however, and you need to deliver new information to them in an effective manner.

For example, if you are presenting a paper at a technical conference, you can assume the audience shares the same educational background and has the same baseline of knowledge. You can connect with the audience by telling anecdotes that are typical of the profession or create comparisons to well-known projects or people.

Another different example arises if your firm is chasing a big new project and you are asked to address a selection committee. In this case, you and your audience have the same baseline information about the project, so you can focus on highlighting why your team is the best for the project and what relevant experience your team brings to the table. A well-prepared and -delivered presentation to a selection committee might just be the boost needed to win the project.

The Public

The general public is definitely the hardest audience to address. When you give a presentation to the public, you are never sure who is sitting in the audience. Your listeners may have a vast array of knowledge and deep understanding about your topic. Or they may know nothing or, worst of all, everything they know is incorrect. The best approach is to keep things as simple as possible. Consider interacting with the audience to get feedback on how they are receiving your message. For instance, tell them you are willing to take questions along the way and then pause in certain spots to look out over the audience to determine how they are reacting to the speech. Encourage them to ask for definitions of any terms they do not know.

Remember that when you are addressing local elected officials, members of the public are also listening to your presentation. Local elected officials may not have the best grasp of engineering issues, but they probably understand them better than the average citizen. Be careful to use terms the general public knows, and do not be afraid to throw in an illustrative story. Elected officials may be your primary audience, but others at the hearing need to understand the project after the presentation.

Here is a brief example of how hard it can be to deal with a public audience. Most of us love our cars and hold strong beliefs in our ability to be great drivers. It is hard to overestimate the lengths to which people will go to make sure they can use their cars, even if there are reasonable alternatives to mobility. People will pay just about any price for gas and sit in any amount of traffic just so they can park their cars at the office while they work all day. Moreover, 90% of us believe that we are above-average drivers. So it should not be surprising that most people firmly believe that they are traffic engineers. They will express opinions about what should be done to make our streets safer, quicker, slower, wider, or smoother. They care little about what a traffic study might conclude. Forget the experts—their opinion is all that really matters. If people listened to them, then traffic problems would be solved. Money would not be an object: the city should be happy to pay whatever it costs to implement their ideas.

Getting People to Listen to Experts

As you can see from our discussion of an audience of the public, it can be a struggle to get them to agree that the speaker is an expert when they do not understand the need for expertise. Communicating with an audience as a professional requires you to find commonality with them. Determine where you and the audience agree, highlight the connection, and move forward explaining how your skill set will make things better. Experience and training save time and money because they avoid future mistakes. You cannot cram it down the audience's throat, but you can tell them why engineering know-how is valuable.

If you are speaking about a road upgrade, tell them how much you do not like sitting in traffic and are working to alleviate everyone's problem. That may make for the perfect story to begin your presentation. Use that story to relate your goal to make traffic move more efficiently and that it may not be as easy as it seems. Talk about obvious alternatives that are not viable once a knowledgeable engineer assesses the situation and determines the problems with the apparent solutions.

Do not shy away from both sides of an argument. If the audience has valid concerns, acknowledge and address them by providing pros and cons about what is proposed. Just telling them you are the engineer and they should trust your judgment will not win over an audience with questions and concerns. If you cover up or ignore their apprehensions, they will not find you a credible speaker or knowledgeable expert.

Facts or numbers alone will not win over a public audience. They can be the foundation for why your engineering decisions are correct, and you will need to present them as the basis for your choices. However, you still need to provide the audience tangible comparisons to help them understand why a decision was made a certain way. Conveying your expertise and the benefits of that knowledge to the public can be difficult, but it's worth the hassle.

The Planning Checklist (Fig. 2-1) should be used to help you determine the audience type to which you are speaking.

* * *

As Vivian prepares to make her presentation to the city council, she spends some time thinking about her audience type. She will be speaking primarily to seven locally elected officials. Four of them are men and three are women. Five have college degrees and two do not. Two of the council members serve on the public works committee and five do not. One council member is a retired architect, and the rest of the council members have occupations that are distant from engineering or construction.

Vivian's secondary audience is the public that will be attending the meeting. The project neighbors will be business owners along the street worried about access to their stores. They will be in their 40s and 50s, with more men than women. Most of them have had their businesses on the street for several years and are successful at what they do. None of the business owners are involved in the construction or engineering field.

Case Study: Complete Streets

One of the basic tenets of good community planning is keeping busy streets out of residential neighborhoods. This is a fairly easy principle to understand, as few people enjoy living directly on a busy street. They would much rather live on a street with a few cars driving by at a slow speed. However, as a community grows, sometimes a lightly traveled street becomes one that is heavily used.

During my time on the city council, I became aware of just such a street. What was once a quiet residential street had turned into a busy collector street. The street was in my ward, so I got involved in working with the neighbors to reduce the traffic, or at least the speed, on the street. It appeared that the amount of traffic was increasing because an adjacent arterial road needed to be widened. The arterial widening slowed when it ran into right-of-way acquisition problems, and the resulting congestion made it undesirable to drive on the arterial. Consequently, an old neighborhood street was taking on more of the burden of traffic moving east–west in the city.

A couple of neighborhood meetings were organized to figure out how to reduce the speed of traffic on the neighborhood road. Drivers using the street to get across town tended to increase their speed because they did not see it as residential. These drivers did not think of themselves as cutting through a neighborhood, but as taking a through-road to their destinations.

The neighborhood quickly learned that two different city departments could be called upon to help curb the traffic speed. Police departments can slow speeders by increasing patrols. Public works departments can be asked to configure a road so that it makes drivers feel like they need to slow down.

Things get interesting when these two departments try to hash out who should be in charge of slowing the speed of drivers. The police department will tell you that they are not terribly effective at slowing speeders. If they post a car on the road, people will slow down until the car is no longer there. Once the car is gone, people speed up again.

With limited resources, they can't put a cop on every collector street in town. The police department concludes that the only real way to slow down speeders is to change the road configuration. The public works department will tell you that they can configure the road in whichever way you want, but without traffic enforcement, speeders tend to find a way to speed. They do acknowledge that some effective measures can be taken to configure a street to slow traffic.

For my neighborhood street, the public works department did agree to look at some options for configuring the road to slow the traffic. One problem with this particular road was that it had been built up over many years. Thus, the road width is not consistent from one end to the other. On the east end, the road is quite narrow, and it broadens to a fairly wide road at the west end.

Among the options proposed by the public works department, the most expensive was reconstructing the road to a consistent width. The city had no money available for this road construction, and the considerable cost would have to be paid by the residents on the street. At a more moderate cost, speed bumps or other speed-control devices could be added, with the cost also borne by the neighbors.

The final option was to restripe the road with narrow lane widths at a nominal cost to the city and no cost to the residents. Because of the varying widths along the road length, the only way to restripe the street with the narrow traffic lanes was to add a bike lane for the road length. The bike lanes complemented the narrowed lane widths, using the full road width while giving the impression of a narrow driving lane. Based on the options available, it was the cheapest, quickest, and most reasonable way to slow traffic on the street.

The public works department set up a meeting with the neighbors to discuss restriping. The meeting began with an engineer giving a presentation on the alternatives for slowing traffic on the street. Once the presentation was over, though, and the public began asking questions, it became clear the engineer had not effectively communicated with the audience. The questions revolved around bike lanes and the perceived agenda of probicycle groups instead of focusing on the goal of slowing traffic on the street.

The audience for this presentation was a difficult one. About half of the audience consisted of a group of people who were strong advocates for bike lanes and showed up to support bike lane creation. The other half of the audience included the neighbors who lived on the street. These neighbors were not supportive of the bike lanes for reasons ranging from a loss of parking spaces to danger for people using the bike lanes. So 50% were there to support one objective, and 50% were there to oppose the same objective. However, everyone in the audience was supportive of the objective of slowing traffic on that street. A prepared speaker might have found a way to communicate with both groups by emphasizing the shared objective.

Unfortunately, in this case, the meeting quickly turned into a fight about bike lanes. The stated goal of the meeting—slowing traffic on the street—was barely mentioned. The meeting became an unproductive debate on a topic tangential to the problem to be solved. The result: no progress was made on slowing traffic on the street.

In this example, the engineers did the best they could to find a solution that was quick and cost-effective. However, they did not spend enough time understanding their audience to realize they had to satisfy two different groups that evening: the neighbors and the advocates of bike lanes. A speech that clearly covered the public works department objectives was not enough to convince an audience with a different set of objectives.

Admittedly, this audience was particularly difficult. However, by not being prepared for the diverging objectives, the engineer allowed the meeting to run off the rails into a fruitless evening. A speaker better prepared for this audience might have been able to get a better outcome from the meeting.

Assessing the Setting

As Tom and Vivian discuss her upcoming speech, Tom asks her how many times she has attended a city council meeting. Vivian says only once. Tom asks her to describe the room, and Vivian cannot recall many of the room details. Since the meeting room is a few minutes away, Tom encourages her to stop by the council chambers before her presentation. He knows that this step will help her understand how the room is arranged, where she will be speaking, and where the audience will be sitting and will lessen the apprehension she has about speaking. He also suggests that she contact the city clerk to discuss what speaking aids she will need to bring on the evening of the presentation. He recalls a computer and projector set-up, so she could simply put the presentation on a flash drive and load the file on the computer in the council chambers.

Whenever possible, it's a good idea to scope out the physical attributes of the place where you will be speaking. You can then tailor your presentation to match the space. You can also try to anticipate what might go wrong, so you can have a backup plan in place.

Physical Attributes

Any location or room has attributes that contribute to the success or failure of a presentation, so knowing the physical setting for your speech can help exploit the positives and mitigate the negatives. For instance, you want to know how the seating is arranged. Will you be in the front of the room behind a lectern addressing a large audience? Will the audience be sitting at chairs with a table in front of them, and will they have a place to write? Will you be at a podium addressing the local government council members with an audience behind you? Will you be able to change the seating arrangement if it does not work well with the presentation?

Sometimes you have control over how the room is arranged, but often you do not. See if you can discover the answers to these questions:

- Are the acoustics in the room good or bad?
- How will the chairs be arranged in the room?
- Where will the audience be located?
- Where will the podium be located?
- Will a microphone be used?
- Is there a projector I can use with my computer?
- Is there a screen I can use with the projector?
- Is there a table available to set up my computer?
- What type of computer (Mac or PC) can I use?
- If I want to bring charts or other visual aids, is there a place to display them?
- Are electric plugs available?
- Is the Internet connection wireless or Ethernet?
- Who is responsible for ensuring that the computer, projector, and remote control are working?

A short reconnaissance before the meeting is often helpful to get a feel for the room. You can see the layout of the tables and chairs and where the audience will sit. Find out if there will be a podium to stand behind or if you will be sitting while making the presentation.

Look around the room to determine where would be a good place to put up any visual aids, if there will be a projector available, or if you will need to bring one. Determine whether you can load the presentation on the computer before the meeting starts. If you plan to use a microphone, see if there is a time to practice with it. It is pretty easy to find out this information, either through a phone call or stopping by the room 15 or 20 minutes early.

Arriving early for your presentation should give you a chance to set up and check that the computer, projector, screen, and any remotes you want to use in your presentation are working. You can familiarize yourself with aspects of the technical pieces of the presentation you are using. If you are not using your own equipment, load your presentation on their computer before the meeting starts and give it a test run.

Finding the Space

Every presentation takes place in a space. As the speaker, you are responsible for getting yourself to that space to deliver the speech. A conference room in your office building is not going to take much effort to find. You are familiar with the location and have been in the room many times. A meeting room at a conference center across town or across the country may not be as easy to find.

A road map will lead you to the building, and a building map will lead you to the room. If you are giving an important presentation, consider making a test run to the building and room location. This will give you an estimate of how much time it takes to get there and orient you to where you are going. You will need to find the best place to park and how to get from your car to the meeting room. If you know what time you are speaking, you have the ability to plan the trip from the office, or home, or hotel to the presentation space. This type of advance scouting will help dampen speaking anxiety because you have one less thing to worry about. Once you arrive at the presentation room, step inside to do an evaluation of the space.

Evaluating the Space

Advance scouting of the presentation room should include spending several minutes inside the room looking for key presentation elements. If possible, you should visit the space several days in advance of the speech. If you can't do that, plan on arriving at the room about

20 to 30 minutes early to peruse the space and set up your presentation. Visual aids and presentation tools should be in place before the speech begins. Try to avoid wasting presentation time setting up the slides or placing pictures on a tripod. Those things should be ready to go before you begin speaking. A quick test of the presentation components (computer, projector, computer file) before the meeting will help ensure that the presentation runs smoothly.

Figs. 2-2, 2-3, and 2-4 show three typical spaces set up for public speaking. Each room is a little different, and giving a presentation in them will be slightly different. Let's scrutinize them to find out what is in the space and what is not.

The room in Fig. 2-2 has a ceiling-mounted projector that displays on a white board. A computer is hooked up to the projector (not shown in the picture). The desks are turned toward the center of the room, so addressing the audience may be a little difficult and take a conscious effort to look at both sides of the room. The room is not very big, so no microphone is needed. The audience would be able to see graphics on a flip chart. You cannot see them in the picture, but plug-ins are located throughout the room and wireless Internet is available.

Fig. 2-2. A Typical Room Set Up in a Classroom Style

Fig. 2-3. A Typical Room Set Up in a Boardroom Style

The room in Fig. 2-3 has a projector on a small table and a screen for display. No computer is hooked up to the projector, so the speaker needs to bring one. The tables are in a U shape with the focus on the center of the room, so addressing the audience from near the screen is best for making eye contact. The room is not very big, so no microphone is needed. The audience will be able to see graphics on a flip chart if it is placed near the screen. The room has one set of electrical outlets near the door, and wireless Internet is available.

The room in Fig. 2-4 does not have a projector or computer, so the speaker needs to provide those items. A small table at the front would allow the speaker to set up a projector to use the screen (which is currently rolled up on the wall) for display. The audience will sit in rows of chairs facing the speaker, so it will be easy to maintain eye contact with them. The room is bigger than the first two rooms, so a microphone is probably needed. The audience sitting in the back may not be able to see graphics on a flip chart if it is placed near the screen. The electrical outlets are on the wall at the front, and wireless Internet is not available.

The Planning Checklist (Fig. 2-1) is a good tool for helping you recognize the type of room in which you will be speaking.

* * *

Vivian decides to embark on her scouting trip to the city council chambers to look over the room where she is making her presentation. When she arrives, she begins looking at the room from the perspective of

Fig. 2-4. A Typical Room Set Up in a Theater Style

a speaker. She walks up to where speakers address the council. The seven council members will be seated in front of her behind a long semicircular desk, and the public will be behind her in rows, with staff sitting to the left of the council chairs.

At the podium, she sees a microphone and notices a projector hanging from the ceiling directly in front of her. The projector points in the direction of the wall used to display slides. She looks down to her right at the computer, and it appears to be hooked up to the projector. She sees a power strip below the computer. Her final assessment involves pulling out her phone to see whether there is a wireless Internet signal in the room. Her phone does find a signal, so she knows she will be able to use the Internet during her presentation.

The council meeting starts at 6:30 p.m., so she does a quick schedule to figure out when she will need to leave work to be at the meeting on time. She wants to be in the chambers at 6:10 p.m. to set up her presentation. It takes her 30 to 35 minutes to get from the office to the parking lot during afternoon traffic. Once she is parked, it takes about five minutes to walk from the parking lot to the council chambers. So she needs to leave her office about 5:30 p.m. to allow for travel to the council chambers. Vivian feels good about her plan to arrive and set up the presentation, and it will be one less thing she has to worry about before the presentation.

Developing a Plan B

A well-prepared speaker who has visited the presentation location and looked over the space for key presentation elements should develop one more plan: a plan for how to proceed when things go wrong.

Here is what can happen. The murmur of the crowd begins to quiet as the moderator asks people to settle into their seats for the next presentation. After a brief introduction, the speaker emerges at the front of the room to begin his talk. He is three slides into the speech, just delving into the project history, when the projector light goes out. Although the projector's light is now dim, the look on the speaker's face suggests a deer in the headlights.

We have all been in presentations where the technology did not work. It can happen to anyone. Some speakers are able to deal with it, and some are not. If you are unlucky, it has happened to you. That brings us to Plan B. What happens next depends on your preparation and self-confidence. If you included a Plan B as part of your speech preparation, then you can move on from the mishap with little disruption. You might even impress the audience with your grasp of the subject matter and your ability to carry on without the projector.

So as you prepare, consider a backup plan for a situation where the technology in the room fails. How are you going to give the presentation if the computer dies? What if the projector lightbulb goes out halfway through the presentation?

Start with the basics. The most common equipment used in a modern presentation are a computer, a projector, a projection screen, presentation software, the computer file with the presentation, a remote control, a microphone, and audio speakers. Electric power, Internet access, and laptop batteries should not be overlooked. Which of these items can you do without? Which are essential? Can you provide a backup for the essential items? For instance, can you bring your own laptop, with a variety of cables and a backup battery, "just in case"? A breakdown of any one of these elements can derail your presentation.

Being prepared means knowing what you will do if anything goes wrong, so you can quickly correct your course and resume your presentation.

I'll talk more about practicing your speech in the chapter on delivery; in this context, think about practicing to give the speech without a projector. Or a computer. Or a presentation file.

A good contingency plan has three steps, and Fig. 2-5 can help you organize your own Plan B.

Plan B Checklist

Fill in the blanks and circle the options that apply

If the computer for the presentation does not work, I will

____

____

If the projector for the presentation does not work, I will

____

____

If I get sick or can't speak, I will

____

____

If the file with the presentation does not work, I will

____

____

If the Internet does not work, I will

____

____

I have a backup copy of the file on

Another computer that I can easily access	A memory stick
On a website I can easily access	In an e-mail I can easily access

I expect ____________ people in the audience.

I will bring ____________ handouts for the audience.

Fig. 2-5. Plan B Checklist

The first step is to identify the hazards and determine what can go wrong. These are things like the following:

- A projector bulb goes out and no replacement is available.
- The speaker gets sick and cannot give the presentation.
- The software file is corrupt and will not open.
- The computer battery is dead and no plug-in cord is available.
- The handouts for the speech are forgotten at the office.
- The microphone that allows you to speak to a large conference room is not working.

The second part of Plan B is to determine the risk level. If you are giving a weekly status update on the progress of a project, then there may not be a lot at stake if the computer malfunctions. A projector going out during a presentation to a project selection committee can spell doom if you have no way to complete the presentation. Many of these risks can be mitigated with backup plans for the elements involved in a presentation.

The third part of Plan B is preparing for an emergency. Think about the elements that will be involved in the presentation and how to address a failure. Put a backup file of your presentation on a flash drive, in case you cannot use your computer or the file will not open. If the plan is to use an unfamiliar computer, bring along a backup. Is there a reliable colleague who can give the presentation if you are sick or cannot make the presentation? If the presentation is based on showing certain graphics, do you need to have posters in case the projector breaks down?

The most important thing you can do to mitigate a failure of technology is to know your presentation well enough to give it without the technology. All that technology is just a tool, never a substitute for a well-practiced presentation.

* * *

Vivian's trip to the council chambers allows her to understand the physical setting and how her presentation will be delivered. While she is standing there, she fills out her Plan B checklist. She will need to bring handouts, a flash drive, a computer, and a projector—the computer and projector in case those elements fail during the presentation, a flash drive to place the presentation file on the room's computer, and handouts in case all else fails. She makes a note to herself to talk with Tom about his ability to make the presentation if she were to get sick and could not attend the meeting.

Finally, before she leaves, Vivian spends a few moments behind the podium with her eyes closed visualizing her presentation. She runs through how the presentation will go and what the room will look like when she is speaking. While she is not in the planning and delivery stages of the speech, Vivian takes this opportunity to practice a good habit for speakers. When she practices the speech in the future, she will easily be able to construct the room and setting in her mind because she has been there recently.

She will lessen her anxiety about the speech because she knows she has a backup plan. She does not have to worry about all the things that might go wrong because she has already made a plan in case they do. Having done the scouting and backup planning, Vivian feels more prepared for the presentation and less anxious about speaking to the council. She is planning a good speech, and the planning will give her self-confidence when she does speak.

Case Study: Murphy's Water Tank

Our firm responded to a Request for Proposal from a public water supply system that needed an engineering firm to upgrade its water tank. Our proposal was ranked in the top three, so we were asked to come in for a presentation and interview. We were excited about the opportunity to work with a new client, and we did a great job of preparing. We got started early and developed a good set of objectives and slides. We practiced several times as a team before the interview and went into the meeting well prepared.

What we were not ready for was everything else with the presentation going haywire. We had 20 minutes to give our presentation, and we ended up spending about five minutes of that discussing the project. The rest was taken up with technical difficulties that could have been avoided with a little planning.

We had been told that they would have a projector available, but we needed to supply the computer. When we arrived at the meeting room and started setting up, their projector and our computer did not sync. We tried three times to get them to talk to each other, but with no success. Of course, we did not bring the projector we had practiced with because we trusted that their projector and our computer would easily integrate. That usually happens, but in this case, we could not make it work. If we had added our projector to the items we brought to the interview, we could have easily solved the problem.

One of the members of the committee said we could use his computer that was certain to sync with the projector, so we gave it a try. He hooked up the computer, and sure enough, his screen was projected on the wall in no time. Then he asked us for an electronic copy of the presentation file to load on his computer. We had not brought the presentation on a flash drive because we were using our computer. Turns out that no one at the interview had a flash drive on them. We were able to eventually share the presentation via Dropbox, and it was loaded on his computer and projected on the screen.

At 10 minutes into the interview, we finally have the opening slide on the wall. We begin our presentation, and about four slides in, we realize that the version of the presentation on the wall has changed from the one we used to rehearse. We had updated the slides on the computer we were going to use, but had not updated the file on Dropbox. So we had to plow forward with the old version and describe to the committee the slides and graphics we had inserted in the presentation.

About eight slides into the presentation, his computer announces that it needs to update its software, so it begins the process of shutting down and restarting, with no way to stop the shutdown. Now we no longer had the use of the slide show on the wall. We had not planned for giving the presentation without a computer and projector, so we had not printed out the slides. If we had, we could have at least handed out our slides.

At this point, everyone on the team was very flustered, and the committee members were not focused on hiring us as their engineers. You could tell that they felt bad about all the technical difficulties, but their charity was not going to get us the job. We spent the last three minutes of the interview doing our best to talk about how we would approach the project, but it was not very effective. They asked a couple of questions, and then we packed up our equipment.

We walked out of the room to see the next group of engineers waiting for their turn to present. We hoped their presentation would be as much of a disaster as ours, but that was unlikely to be the case.

To no one's surprise, we were not selected as the engineering firm for the water tank project. We spent a lot of time and effort getting ready and had a good presentation, but the committee was not able to see it.

Types of Meetings

Meetings with presentations come in various shapes and sizes, ranging from formal speeches before large audiences to casual talks with co-workers. Analyzing the characteristics of the meeting can help determine the amount of preparation time needed, the focus of the audience and its objectives, knowing which visual aids to develop, and fitting the presentation into the allotted time.

Meetings take place for a variety of purposes, with a variety of structures. Some common meeting types at which engineers are likely to speak include project, staff, sales, conference, collaborative, kick-off, community, committee, status update, proposal, selection committee, online, problem-solving, decision-making, informational, and evaluation.

Each type of meeting has a unique set of characteristics that define how it is conducted.

As you prepare your presentation, think about the type of meeting and how to tailor the presentation to fit the meeting characteristics. A few examples of different types of meetings are described below.

Local Government Meetings

Local governments use formal meetings to conduct their official business. They are structured meetings with an established protocol and a presiding officer (such as the mayor). These meetings have a prescribed agenda that is followed during the course of the meeting. Speakers are often given a strict time frame for completing their presentations, and questions are taken at the end. Speakers are encouraged or required to keep presentations to the set time limit. The mayor or council chair conducts the meeting following the agenda and Robert's Rules of Order. Typically, speakers give their presentations, answer questions, and return to their seats when finished. They may be called on later to answer other questions that arise. After asking questions of speakers or staff members, the council members will debate the issue. All decisions will be made based on the votes of the governing body members, and the votes during the meeting will be recorded for the public record.

Conference Presentations

Conference sessions provide speakers a chance to present their professional work to an audience of peers. These sessions allow speakers to highlight their unique knowledge or specialized experience on projects. The audience will share sensibilities and education with the speaker, so some level of professional jargon is acceptable. Introducing the speaker is common, and a prepared speaker will have a short biography ready to give to the person doing the introduction.

At conference sessions, the audience can respond very differently according to the time of day. Morning sessions, for instance, usually bring an attentive audience, while an afternoon session can bring a sleepy audience. Adding a bit of extra pep or activity to an afternoon conference session can improve your presentation. Be sure to arrive at the room with your session before it is scheduled to begin, even if you are not the first speaker in the session. That gives you plenty of time to set up and make sure all the components (computer, projector, presentation file) are ready to go when it is your turn to speak.

Staff Report on Safety

Developing a culture of safety in an organization means discussing those topics on a regular basis. An organization may rotate speakers giving presentations on safety at the weekly staff meeting. Presenters are given a topic or asked to speak on a subject of their own choosing. The objective is to impart safety knowledge and procedures to an audience of coworkers, who have listened to similar safety talks. Audience members are more likely to share the speaker's sensibility or technical background. As the speaker, you may be challenged to make your presentation sound fresh and stand out. Most likely, you will be speaking in a familiar conference room, so setting up the computer and projector should be easy. And you will know the room setup, acoustics, and where the audience will be sitting.

Government Committee Meetings

Meetings of governmental committees can be highly structured, but they tend to be less controlled and stressful than, say, a local council meeting. The structure of committee meetings is more casual, and they often involve several members of a staff or an interdisciplinary team. These meetings usually have an informational purpose, and the decisions made at them are more mundane than critical. These meetings encourage in-depth discussions in a way that formal meetings do not. Committee members will engage the speaker during the presentation, ask questions, and challenge assumptions.

An effective presentation allows committee members to get a deeper understanding of an issue before it is brought to the formal meeting for a decisive vote. Government committees are often organized around topics important to the local government. They are likely to be called the Public Safety Committee, the Budget Committee, the Planning Committee, or something along those lines. Committee names designate the duty of that committee to study an issue in depth and then, if the committee desires, bring that topic for action by the full council. Speaking at a committee meeting may be less official than a formal meeting, but it needs to be done well. Find out before the meeting the protocol of the committee and the best way to deliver a presentation.

Sales Meeting

Engineers involved with representing and selling industrial or municipal equipment are often making presentations about the equipment or products they represent. Typically, they are meeting with other engineers who want to learn about their pumps, valves, geotechnical fabric, PVC pipe, or any of the product lines or upgrades the sales engineer represents. Sitting down with a potential customer and having a discussion is a good way to get the message out about a product. The setting is less formal, and people involved with the meeting normally sit around a conference room table. The presentation itself tends to be less formal, and the speaker is more frequently interrupted by questions and discussion from the listeners.

These presentations should highlight the product's capabilities and demonstrate why your product works well and where it should be used. You should anticipate the client's needs and objectives, emphasize where you can be helpful, demonstrate successful projects, and allow plenty of opportunity for discussion. If you are invited to address a small group meeting, be sure to ask how the host would like the information to be presented. Are slides required? Would an informal discussion over a cup of coffee be better? With the smaller group, you should have an easier time determining their objectives. A small audience presentation offers a chance to make a clear connection and communicate about the things important to them. Time lines for these meetings are less rigid and can range from five minutes to an hour.

Proposal Meeting

In selecting an engineering firm for a project, it is common practice for the client to invite the top two or three firms for an interview. During the interview, each firm will make a presentation and answer questions from the selection committee. These meetings usually occur in smaller rooms with only a handful of people in the audience. More often than not, the meeting room is an unknown, so scouting and planning are advantageous. The audience is especially important because they will be deciding the firm chosen for the work. Most are intimate, with fewer than 10 people listening to the presentation.

One Topic, Many Meetings

One final note on giving a presentation at different meetings. If you are working on a project of interest to many audiences, you may have to give the same presentation, or a similar presentation, many times. The project may have several phases that require presentations at each step. Each time you give the presentation, evaluate the category and type of meeting, and adjust the presentation to fit the meeting. You may become tired of giving the same presentation over and over, but remember, the more effective communication occurs on the project, the more likely that project is to be successful.

This chapter lists just a small sampling of meeting types, but it shows that meetings have unique characteristics and that a well-prepared speaker will consider the type of meeting at which he or she is presenting.

* * *

It does not take Vivian any effort to identify the meeting at which she will be presenting. It is a local government meeting with the public attending. As she wraps up the planning portion of her presentation, Tom lets her know that she is heading in the right direction. He says her objectives are well defined and appropriate for the presentation. They will serve as the foundation of her speech to inform the council about progress and next steps for the road project. Vivian has discussed the audience characteristics with Tom and visited the room where she will speak to them. He agrees to be the backup speaker if, for some reason, she cannot make the presentation. She is well on her way to a good presentation. The next step is to begin the design of her presentation.

Chapter 3

Design

Design is not what it looks and feels like. Design is how it works.
—*Steve Jobs*

This chapter on speech design covers the components of a well-designed speech, with some help from Vivian and Tom, as she gets ready for her big presentation.

- You'll learn about the important elements that go into developing a good outline that matches the speech objectives developed during preparation.
- You'll weigh the advantages of preparing a word-for-word speech draft versus using a simpler talking-points presentation.
- You'll find out how to plan for visual aids, including tips for that all-important slide deck.

The previous chapter on speech preparation laid the groundwork for creating a presentation that will allow you to connect and communicate with your audience. All of the preparation presented in that chapter will help you to develop a speech that furthers that agenda. A good design is made up of four elements: content, outline, talking points, and visual aids. I will cover them all, but I begin by discussing how you take all the information gathered during preparation—objectives, audience, setting, meeting type—add the content, and create a presentation outline. Tom and Vivian provide insight into this process as they keep working on her presentation.

After a speech is outlined, the outline is used to transform it into a deliverable presentation. I review the options for moving your speech from an outline into a presentation. Finally, I examine one of the most important tools used during a presentation: visual aids. A plethora of visual aids is available, but the most common is presentation software. This chapter provides advice on how to develop visual aids that improve communication.

Three case studies are included in the chapter to illustrate how to put speech design into practice. One case study shows how to outline a TEDx speech and provides a template that you may use to outline a presentation. A second case study demonstrates how to transform an outline into a three-minute speech. This case study includes three examples. The third case study is a cautionary tale of a great engineer who did not effectively use his visual aids and ended up inflicting death by PowerPoint on his audience. Throughout the chapter, Tom and Vivian discuss how she needs to move from planning to design.

Moving from Planning to Design

Over several days, Vivian works hard to prepare her speech. She develops three main objectives, spends some time brushing up on the council members' backgrounds, and investigates the room where she will be speaking. She feels a lot less dread when the tinny bell once again signifies an incoming e-mail, this one with the subject line: "How's the city council presentation coming?" Follow-up e-mails are standard operating procedure for Tom, who likes to check up on conversations a few days after they happen. Vivian opens the e-mail to find this:

> Vivian,
>
> Just wanted to inquire about our conversation a couple of days ago on the presentation to the City Council. How are things coming along? Please stop by my office and give me a brief update.
>
> Thanks
>
> Tom

Vivian needs a cup of coffee, so she wanders down to Tom's office and pops her head in. She updates Tom on her progress, and he agrees it is a good start to the speech. Then Tom asks her how she is going to build her bridge. Vivian is confused—she doesn't have any bridge projects at the moment. Tom explains that she has done a great job of planning her speech, but now it is time to build the speech structure. Like a bridge, each speech needs a structure to give it support and direction.

A bridge is a good metaphor for a speech outline. You need to get from Point A (the audience does not have the knowledge you have) to Point B (the audience has that knowledge because you communicated it effectively). Between Point A and Point B stretches an empty space that needs to be spanned. Among the elements that allow a bridge to span the empty space is the structure that supports the deck. A speech needs the same type of structure to support it. But rather than concrete and steel beams, a speech relies on content, an outline, talking points, and visual aids.

Developing an Outline

To get started with your outline, pull together everything you know about the presentation and put it in some kind of order. This content will allow you to develop a speech outline looking at the big picture before polishing the details. You will be taking into account the objectives you identified in Chapter 2, Planning. The audience, speech objectives, the time allotted, and your setting are important factors in shaping the outline. But first, you need to know *what* you are trying to communicate.

Collecting Material

Although most of this book is about the "how" of public speaking, in this section I am going to focus on "what" you are saying. You can call this your message or content. It is the presentation substance and the information you want to convey to the audience. Because you are, most likely, an engineer, your material is probably going to be technical. In some cases, additional research or information collection is necessary for the design of your presentation.

The first step is to determine how much you, the speaker, know about the topic. Chances are you know a lot about it. If you have been asked to speak about a project you are working on or an element of engineering with which you have experience, you will be quite knowledgeable about the subject. That is what should be in the body of the speech. An informed speaker should brush up on the topic, but not a lot of research is needed to figure out "what" will be in the body of the speech.

However, there may be times when you do not have all the information you need about the presentation topic and you need to collect it. This can happen when you are part of a team that is working on a project and you have familiarity with one aspect of the project, but need to reach out to others to get the full project picture. In that case, you will have to do some research by talking to team members and finding out what they know, or what information they have that you can include in the presentation.

There may also be times when you do not know much about a subject on which you have been asked to present. Therefore, you are going to spend time researching the topic and visiting with others about it. This is often a blessing because by educating yourself on a subject you will be surprised how much you learn. The exercise of gathering material and using it as part of a presentation will ensure that you learn a good deal about the issue.

Next, using the speech planning tools as a guideline, determine the relevant information to be included. You probably know a whole lot about the subject, but that does not mean you need to pass along all of that knowledge to the audience. Spend some time planning how to whittle your vast knowledge of the subject down to what is important. It is valuable to match the material in the presentation to what is known about the audience and the speech objectives. If you have to do research on the presentation, you will find all kinds of material relevant to the topic. Do not throw it all into the slide deck. Instead, weed out the material that is not relevant so that your speech is not confusing, illogical, or boring.

It will be necessary to provide context to the speech. The context will be based on the project history, the appropriate amount of technical detail, and the audience. An audience filled with engineers will be interested in the project design standards, but an audience of city council members will not. If the audience will not be concerned with design standards, do not spend time researching them for the presentation.

Audience research also provides insight on the questions they tend to ask. If you can anticipate some of the questions, you can gather the information you need to answer them. Approach colleagues who have spoken to a similar group to find out the questions they received during their presentations. Ask them what they wished they had been ready to answer but were not.

Before moving on to outline the speech, be sure you have collected all of the information you need and then decide how much must go in your presentation. This step will be helpful to creating a logical speech outline.

* * *

As Vivian jumps into the design of her speech, she feels confident in her expertise about the project. She knows the technical aspects and the current state of the project. Where she needs help with the presentation is some of the initial project history. She was not involved in the team that did the preliminary work, but she can rely on Tom to provide that information. Since she has not presented to the council previously, she has a tough time anticipating the questions they may ask. She asks Tom about past project history and the types of questions he has gotten previously. Tom recommends she reach out to the public works engineer assigned to the project and also ask him what types of questions to expect. He reassures her that he will come to the meeting and sit in the back of the audience just in case there are questions she cannot answer. Finally, Vivian needs to touch base with the geotechnical subcontractor to find out where they are in completing their work and their initial impressions of the project.

Building an Outline

Essential to almost every construction project is a well-designed set of engineering drawings. They provide all of the people working on the project a clear, coherent, and focused blueprint for how to complete a complicated construction project. The role of an engineer is to design those plans so that the complex portion of the project can proceed with as much ease as possible. A set of plan sheets reflects key logical and fundamental project elements.

The outline of your speech will provide those same elements to the presentation. It will make your speech logical, coherent, and focused. At a basic level, the outline of a speech looks like this:

Introduction
Body
Conclusion

The introduction is an opportunity to let the audience know who you are and why you are making the presentation. It allows you to tell them why you are speaking and preview what you want them to learn during the speech. The body is where you provide the audience with the information. It is the presentation's core and your chance to communicate the main points of the message. The conclusion is the wrap-up of the presentation and where you summarize the key points you are communicating to the audience.

Engineering plans are rarely rudimentary, and neither is a well-designed presentation. The next outline layer is to add the presentation objectives. Adding them will update the outline to look like this:

Introduction
Body
    Objective 1
    Objective 2
    Objective 3
Conclusion

The outline is still basic at this point, but you should be able to see how to build on this basic outline, adding more detail, until you have an outline that serves as a presentation design. The best way to demonstrate going from a basic outline to a completed outline is to use a few example speeches to explain the process. Let's consider three different speeches and discuss how to outline those speeches.

The first is a presentation at a technical conference. These are generally informative speeches and are designed to educate the audience. Your goal is to inform an audience of peers about a topic in which they have shown interest.

The second is a speech that highlights some project decision points. This type of speech contains elements of an informative speech, but also some elements of a persuasive speech. You are going to present options to the audience and information that supports your best judgment on how the decision should be made.

In the third speech, you are making a recommendation to the audience. In this case, along with elements of an informative speech, you are setting up options to persuade the audience to agree with your recommendations.

For the first example, being selected to speak at a technical conference is a great honor. It means that you were involved in a unique project or groundbreaking research. The conference organizers want you to speak to other engineers, or technically minded people, attending the conference. They want you to pass along your knowledge to a group of colleagues. This type of presentation is largely informative. In this example, the speaker has time to address four objectives. Here is a suggested speech outline for a distinctive project:

Introduction
  Introduce self and preview the presentation
Body
  Objective 1—Project History
    How and why this project was undertaken
    Who funded; who designed; who built
  Objective 2—Exceptionalism
    What made the project time line difficult?
    What makes the engineering design and project process rare?
    What geographic setting and elements provided major obstacles?
  Objective 3—Teamwork
    How did the team overcome these obstacles?
    What unique solutions were offered because of teamwork?
  Objective 4—Lessons Learned
    What did the team do right?
    What would the team do differently?
Conclusion
  Presentation review
  How the project is working since completion

Vivian's speech to the city council is a good example of a project decision point speech, the second example. The presentation will provide elements of an informative speech (Objective 1—Deliver preliminary construction cost estimate, discuss funding and time frame, and Objective 2—Graphically show the beginning and ending of the project and what streets and businesses will be affected) and elements of a persuasive speech where she lays out choices the city council needs to make (Objective 3—Cover the next steps for the project and decision points). Her outline starts with an introduction of herself, a presentation preview, and brief project history and overview. The speech concludes with a presentation summary and a reminder of the decisions before the council:

Introduction
  Introduce self and preview the presentation
  History and project overview, including milestones already passed

Body
- Objective 1—Budget and Time Line
  - Deliver preliminary construction cost estimate
  - Discuss funding and time frame
- Objective 2—Project Limits
  - Graphically show the beginning and ending of the project
  - Discuss what streets and businesses will be affected
- Objective 3—Choices
  - Cover the next steps for the project
  - Discuss key decision points

Conclusion
- Presentation Summary
- List of key decisions points to allow the project to proceed

The third example speech is to persuade a community that a bridge is in need of replacement. This speech has informative elements, but more of it focuses on building a case for a particular conclusion. The speech has three objectives based on three reasons why the bridge should be replaced. An outline of this type of speech might look like this:

Introduction
- Introduce self and preview the presentation
- Role the bridge plays in the transportation system
- History on how long there has been talk of bridge replacement

Body
- Objective 1—Safety
  - Age of bridge and design life of bridge
  - Deficiencies of this bridge
  - How deficiencies can lead to decreased level of service or bridge collapse
- Objective 2—Resiliency
  - How many cars and trucks travel across bridge daily and/or yearly
  - Other ways to cross the river
  - Is bridge a vital link or nonessential?
  - How long it takes to detour around the bridge
- Objective 3—Best Alternative
  - Preferred alternative for bridge replacement
  - Other alternatives for bridge replacement

Conclusion
Presentation Summary
Reasons the preferred alternative is the best option

These three speeches and outlines are example projects. Your project will be different, so structure your outline to meet the presentation's objectives. You can add or subtract objectives to suit the situation, but it is vital for a good presentation to have an outline.

As you build your outline, keep in mind that the amount of time allotted for the speech will influence its outline. There is a big difference in what can be covered in a five-minute speech versus a 90-minute speech. You should be aware of the time limitations and adjust your outline, and objectives, accordingly. If you do not know how much time you have, find out (see Chapter 2). It does no good to prepare a 30-minute speech and find out when you get to the site that you only have 10 minutes. It is just as bad to prepare a 15-minute speech and find out that you are expected to speak for an hour.

Case Study: TEDx Speech

It was a chance encounter late in the evening. I did not go looking for it, but stumbled upon it nonetheless. The look was appealing, the experience would be pleasurable, and the serendipity was perfect. What was it? A chance to speak at a TEDx event.

TEDx events are local TED-like experiences planned and coordinated independently, under an agreement granted by the TED nonprofit media organization. TED is short for Technology, Entertainment, Design, and it began as a conference to share ideas across those fields. Speakers were asked to put together a short speech, less than 18 minutes, which covered a topic they wanted to share with the audience. Over the past 30 years, the TED conference has proven to be extremely popular, and thousands of TED talks have been posted online. In order to expand on the original idea, the TED organizers allow local groups to organize TEDx events where area speakers give a short speech on a topic for which they have a passion.

I do not spend much time on Facebook, but one evening I happened to be looking over my news feed and a friend shared a link to an upcoming TEDx event. Being a fan of TED, I was intrigued by the opportunity to speak at a local TEDx event. I clicked the link to the website, filled out the form, and began the process of becoming a TEDx speaker. It was a wonderful opportunity, and I had a great time planning, designing, and delivering my speech.

The planning portion was relatively painless because I was speaking on a topic I already knew well. It was easy to pick out my objectives and define my audience, and I knew the room in which I would speak. It was the design phase that would need some work. Normally, I deliver the speech over the course of an hour, but for TEDx I needed to fit the speech into 18 minutes. The speech outline needed major changes to go from 60 minutes to less than 18 minutes. I began the process of outlining the speech almost from scratch.

During the process of cutting the speech, I developed a table to make the process easier. It allowed me to pair the Introduction, Objectives, and Conclusion of the speech with the slides I wanted to use and the time I had. Fig. 3-1 shows the speech outline after I removed large sections of the presentation that needed to be jettisoned to get under the time limit.

The table provides one example of how you may want to develop a speech outline. It would help organize the speech and supply a road map for developing and delivering a presentation. Each column describes an aspect of the speech outline, and the row shows how they tie together. Fig. 3-2 is what Vivian's speech would look like if she used this template.

From Outline to Speech

At this point, Vivian has an outline showing how she can incorporate her objectives into a presentation. She has also allocated an amount of time for each section, and the slides she wants to use, which will help her reach the right level of detail for the presentation. Now her task is to make the transition from outline to a speech that is ready for delivery.

Talking Points versus Word for Word

Having a completed outline means it is decision time. You are at a fork in the road in the speech design process. There are two schools of thought on moving your speech from outline to ready-to-deliver. One option is to develop a series of talking points to guide you through the speech. The other option is to write down the speech word for word and read it to the audience.

Script	Slide	Time
Introduction—A Chance for a Week Away I distinctly remember returning to high school for my sophomore year, walking through the cafeteria, seeing the excitement all around with the start of a new school year and thinking to myself—I don't want to be here. So I would jump at any chance to get away, especially for a week at the Montana Legislature.	None	Total: 1.5 minutes Begin: 0:00 End: 1:30
Introduction—Paging at the Legislature Enjoying the opportunity to be around the legislative session. Having important people, legislators, spend time with me to discuss what was going on that week. It sparked a passion in me that I have followed ever since. And it led to serving on the Billings City Council.	Outline of Montana and Arrows	Total: 1.5 minutes Begin: 1:30 End: 3:00
Objective 1—Politics and Change Following my passion for politics led to being a student of politicians and elections. The most common type of election is a politician promising he or she is going to change something. And then he or she gets elected and nothing happens. After serving as an elected official, I began to study change because I was surprised at how hard it was to make a change. One of the barriers to change is much more hidden and is often a silent killer of change. That is our culture. Let me give you an example.	None	Total: 2 minutes Begin: 3:00 End: 5:00
Objective 2—Switching Your Diet The American diet based on meats and plants is a pretty inefficient system. A much more efficient system is an insect-based diet. Arguments for the diet.	Picture of a cow Plated bugs Bug market Lobster	Total: 10 minutes Begin: 5:00 End: 15:00
Conclusion—Culture Silently Kills Change In so many cases, culture is the silent killer of change. That is my idea worth spreading. It is a barrier that cannot be overcome if it is unknown. It is an invisible fence.	None	Total: 1.5 minutes Begin: 15:00 End: 16:30

Fig. 3-1. TED Talk Outline

The first thing to keep in mind when making this choice is your comfort level with speaking and the topic. An experienced speaker discussing a topic she knows well is going to use talking points. Talking points are designed to prompt a speaker to cover speech aspects, but they do not provide a verbatim speech. The speaker talks about each portion from the knowledge she has gained through several speech rehearsals. A less skilled speaker covering a topic he does not know well will be more

Script	Slide	Time
Introduction		Total: 1 minute
Introduce self and preview the presentation	Opening	
History and project overview	Summary	Begin: 0:00
		End: 1:00
Objective 1		Total: 1.5 minutes
Budget and Time Line	Current budget	
Deliver preliminary construction cost estimate	Current time line	Begin: 1:00
Discuss funding and time frame	Current funding	End: 2:30
Objective 2		Total: 1.5 minutes
Project Limits	Project limits	
Graphically show project beginning and end		Begin: 2:30
Discuss streets and businesses that will be affected		End: 4:00
Objective 3		Total: 1.5 minutes
Choices	Next steps	
Cover the next steps for the project	Decision points	Begin: 4:00
Discuss key decision points		End: 5:30
Conclusion		Total: 1 minute
Presentation Summary	Decision points	
List of key decisions points to allow project to proceed		Begin: 5:30
		End: 6:30

Fig. 3-2. Vivian's Outline

inclined to write out the speech, or at least large portions of the speech, so that he can read the speech word for word to the audience.

I want to advise against believing you can avoid a decision by writing out your speech, memorizing it, and delivering your speech from memory. When speakers try to memorize their speeches, they become *more* anxious about speaking. They know they are not going to be able to remember every single word, so they become stressed. Almost no one can memorize a speech word for word. Don't set yourself up to be anxious and fearful about speaking before you have even begun. Stick to the choice between delivering from talking points or reading the complete speech.

Even though you cannot memorize the speech word for word, you can remember the highlights or talking points. Put those talking points on a series of cards to prompt you while speaking. After a glance at the card, you can look up, speak, and engage with the audience. You can also use your slide deck to remember the talking points. When you present using slides with pictures or statements, seeing each slide should prompt you to remember the talking point associated with the slide and proceed to speak to the audience accordingly. The beauty of using the slides to keep you on track is that you will look like a speaking pro, giving a quality presentation without notes. After covering that element, you move on to the next slide, which will prod you to cover the next topic.

You can use a little of both methods by writing out your speech, then highlighting the key sentences. Take a list of those key sentences with you to the presentation and have them handy. This method allows you to speak extemporaneously, but you can have a fallback if you freeze during the speech.

Speakers using the talking-point method are more natural, at ease, and relatable because they can look over the audience while speaking. The audience is more attentive when they can make eye contact with the speaker, and a speaker looking at the audience can gesture and respond. A talking-points speech lets you move around the room if you like so that you are not stuck behind the podium. It provides more opportunity for audience interaction and aids in communicating with the audience.

A talking-points speech requires several rounds of practice. I talk more about the importance of speech rehearsal in Chapter 4, but be aware that the talking-points approach involves practice time. It will not hurt to practice reading your speech if you choose to write and deliver the speech that way. However, it probably will not involve as many rehearsals.

If you choose to write out the speech, the preparation time involves writing several drafts of the speech. Compose the speech in a way that is similar to the way you talk. Sentences should be short because few people talk using long, complicated sentences. It will also give you a chance to take a breath between sentences. Format the speech so that you do not lose your place as you read, and consider printing the speech with a large font to make it easier to read.

When you write out your speech, you risk losing opportunities to engage the audience. On the other hand, you ensure that you say exactly what you want to say, and do not stumble and fumble through the delivery, or leave anything out. Engage the audience as much as you can by looking up from the speech whenever possible. You can mark a couple of spots you know well enough that you can look away from the paper and toward the audience during delivery. Be sure to find places where you can add changes in voice tone so that you are not reading the speech in a monotone.

Your comfort level and topic knowledge will guide the decision between talking points and a word-for-word approach during speech planning.

Developing Each Section

Vivian has her speech outlined and now needs to enhance it so it is ready to deliver. Her task is to animate her outline into a speech, and the steps she takes are discussed in this section. Because she is somewhat confident about speaking, Vivian decides to focus on delivering a good talking-points presentation. Look again at Fig. 3-2 to see her speech outline.

Her speech will begin with an introduction of herself and the topic. Often, the meeting protocol requires speakers to identify themselves and the presentation topic, and that is no different in Vivian's case. She will need to give her name, the organization she works for, the position she holds, her credentials, the project she is speaking about, and why she is giving the presentation. Vivian knows all of this information really well, as most speakers do, so she will use a slide to prompt her to deliver all of this information. She will also have a card at the podium that lists all of the things she should cover so she does not miss anything.

Depending on the meeting format, someone else may give a brief introduction to the audience, such as a staff contact introducing the speaker. In that case, listen to the introduction to ensure all of the introductory bases are covered, and if they are not, take a few moments to shore up any aspects that were not covered. Vivian will not have a staff member introduce her, so she needs to be prepared to cover all of those aspects.

It will be easy for Vivian to shift from the introduction to a history and project overview. She will cover the events that led to the presentation and note any aspects of the project that are already complete. It should be a brief discussion detailing current conditions. It is the starting point for why action is needed on the project, current conditions of the infrastructure, cost of operation and maintenance, design life, and estimated time until failure of the road. She will be telling the audience why the current road is not working well and what the project proposes to fix.

She should not overlook the fact that the audience may not be completely up to speed on the project. Any speaker should be cautious about assuming the audience has even a moderate grasp of a given topic. In Vivian's case, it has been several months since anyone has given the city council a project overview, so she should spend some time familiarizing them with the project. She will use a slide and card to prompt her to discuss these issues, but they are aspects she knows well.

In fact, she probably knows them well enough that she needs to guard against skipping over too much information. Engineers often forget that others have not been living with the project for the past several months and tend to skip a discussion on the project basics. If you, like Vivian, are presenting on a preferred alternative for a road project, discuss how it was determined that the road update was needed in the first place.

If your speech is not on a road project, but instead is covering an upgrade to the wastewater treatment plant, talk about the changes in the statutory requirements for wastewater discharge, discussing the rule that was changed, and what problem it is trying to address. You may also want to touch on the useful life of infrastructure and where that piece of infrastructure is in its life span. Be sure to get the audience up to speed on the project so that they are not asking basic questions when you get into the meatier portion of the presentation.

Vivian is speaking to a local government body, so her project history section will include information on the public meetings that have been held on the project. Those meetings were required as part of the road design process and were used to gather information from interested parties about what they wanted to see from the project. She will inform the council members how many people attended the meetings, when and where the meetings where held, and what was discussed. She also plans to recognize four council members who attended the meetings and thank them for being there.

After the background information, Vivian proceeds to the items she wants to cover in Objective 1. She starts here because she knows that this will be an important audience concern. There are few organizations that have unlimited capital to pay for all of the things they want, and Vivian's audience will be no different. Thus, the city council will be interested in the project budget. Vivian will highlight where the project is located in the city's capital improvement plan and provide a project cost estimate. She will mention that the estimates are based on similar recent construction projects. She will have the key dollar figures on her talking-points card and a slide that covers the project budget.

Vivian will provide the audience with information on the project time frame. She will use a slide to provide a visual and put the important dates on her talking-points cards. She wants to mention when the engineering design needs to be completed, when a contractor will be selected, and when construction will commence. During this slide and talking point, she is going to cover the hurdles that need to be overcome for a successful project. There are still a couple of obstacles that could get in the way of the project budget and time line, so she wants to provide the council members with realistic estimates of how much time, money, and effort will be needed to overcome those barriers.

It will be an easy transition from the discussion of when construction should begin to how the project is expected to be funded. She will use the estimated construction cost along with a breakdown of how the road will be funded. The slide and talking-point card will list each funding source involved and how money will be accessed from those sources.

Vivian is able to move on to her talking points for Objective 2. She is going to use the graphical slide of the project beginning and ending points to prompt her to talk about how the project will affect the surrounding vicinity. She knows the road project is going to affect everyone's favorite neighborhood coffee shop, and she needs to explain why the project starts near the shop and how the project will affect it. It will not be pleasant covering the impact, but ignoring it during the presentation will not make that impact go away. Nothing could be worse for Vivian and the project than a council member or other audience member finding out at a later date the project's negative impact on the character of a neighborhood during construction. Several other commercial businesses will also be affected by the project, so she should have a note card that has them listed.

She is also planning to bring up how other local government services will be affected by the project. The new road will make it easier for the fire department to access this part of town. It will also pave the way for a new emergency communications center near the project, and Vivian will add those talking points to her cards.

The last presentation section will cover Objective 3; Vivian plans to cover where the project is headed. Her slide and talking-points card will list the next steps for the road project. This topic serves as a great lead-in to the discussion of the decisions the council needs to make at the meeting or in the near future. She will clearly lay out what choices need to be made and the options for each choice. It is her responsibility to ensure the audience understands they need to make a decision. They may think that she is giving them a nice project update and not realize that the reason for the presentation is that they need to make a decision. If a decision needs to be made, clearly state that you are waiting for them to make a decision.

To wrap up her presentation, Vivian's final slide will lay out exactly what decisions need to be made and when they need to be made. While that slide is showing, she needs to include any final remarks she wants to make and do a brief summary.

As Fig. 3-2 shows, Vivian is planning on using nine slides during the presentation. She will also have at least that many talking-points cards.

Between the two, they will prompt her through the speech to cover topics she knows pretty well.

Oldies but Goodies

Having listened to a lot of presentations, I can offer two nuggets of advice about turning your outline into a speech. The first is to start the speech with a story, and the second is to cover the main topics three times during the course of the speech. Given the right circumstances, each of these can be effective tools in gaining your audience's attention and ensuring that they hear and understand the important points of your presentation.

A great way to engage your audience is by telling them a story. After a proper introduction, consider some sort of narrative to preface the comments about the presentation's main topic. Often, opening with a story translates into telling a joke, but engineering presentations tend not to lend themselves to a great deal of humor. If you can find a good joke that suits the topic, then please use it. However, if you do not have an amusing anecdote on hand, then try to find an interesting story to get the ball rolling. Instead of a full story, you could open with an interesting fact or surprising statistic that is relevant to the topic.

Using humor or good stories in a speech helps grab the audience's attention by making an emotional connection. They will think that you are an interesting speaker and pay attention to what you are saying longer before tuning out. A good opening to a speech gives the speaker a chance to reach his or her audience before they turn their attention elsewhere. Telling engaging or humorous stories is how you start to build a relationship with the audience. Those stories are a powerful way to motivate and inspire your audience to listen to the information you are sharing.

Finding these stories is not easy. You will have to pay attention to things that are relevant to the topic, grab your attention, and make you stop to think or laugh. Gather these topics through conversations with coworkers, listening to the news, or reading the newspaper. The next time you read or hear something interesting, take an extra second to think about how you might use that in your next speech and how the audience would react to the story.

If you have time to tell a story (say, a minute or two), then use a broad brush to show how the project will improve the community. If you are speaking to the group about a road project, talk about the family time lost by commuters, or the loss of lives from a bad road design, or why investment in a road will be good for local commerce. If you are speaking about an upgrade to a water treatment plant, talk about the health aspects of clean water and the importance of avoiding the spread of waterborne disease or about the value in the drinking water system.

For example, if you are speaking about a road project, you could start the presentation by telling the audience the following:

> Poor road conditions cost the average family $360 per year in additional upkeep and maintenance on their car, and, on average, each commuter spends 45 minutes per day in slow or stopped traffic. Over the course of a year, that adds up to 187.5 hours or about 4.5 weeks of time lost to traffic.

If you are speaking about a water system project, you could start with this:

> It takes approximately 7.5 bottles of water to make a gallon of water. If that bottle of water sells for $1 at the local grocery store, it makes the price of a gallon of water about $7.50. Our city sells a gallon of water for $0.013 per gallon, providing safe drinking water throughout the city for a fraction of what you spend at the grocery store or gas station.

Remember that engineers have a compelling story to tell about how they improve the lives of those around them. Do not be shy about telling that story and using it to open a presentation. Try to find some aspect of the project you are passionate about and expound on that aspect for a few moments in the presentation's opening. Keep the length of the story proportional to the length of speaking time, but do not be afraid to grab the attention of the audience with a narrative about how this project will make things better.

The second piece of advice about giving a speech is this: "Tell them what you are going to tell them; tell them; and tell them what you told them." The idea behind this is to give the audience three chances to hear the message. Begin the presentation by going over what you want to cover in the speech. This is often done with a single slide that outlines what the presentation will cover.

Then present the main topics of the speech, working hard to communicate with them what it is you want them to know. After you have completed the body of the speech, take another moment to recap what was presented in the speech. Again use one slide, often with the same outline as the first slide, summarizing the main points. When reviewing, use a phrase such as: "If there is one thing I want you to understand after listening to my speech, it is that the bridge is in need of replacement."

One of the reasons this advice has been around for so long is that there is a great correlation between the approach and the basic speech outline. The introduction becomes "Tell them what you are going to tell them." The body is the part where you tell them. And in the conclusion, you tell them what you told them.

Adding the intro and conclusion to a speech should be easy. Working a story into a speech will help with audience communication. However, if it is a struggle to find the perfect anecdote to open the speech and it is hampering your efforts to build an outline, then skip it. If you have a tight time frame for the speech, then you should consider skipping the preview and recap of the speech and focus on the main presentation message.

Keep in mind that it is easy to become long-winded during presentations. You may have a great passion for the topic and want to share your enthusiasm for the project. Most audiences will not share your passion for technical details, such as the chemical reactions behind the upgrade at the wastewater treatment plant. Give them the information they need, and do not risk losing their attention with too many technical details about the project just because you really, really enjoy discussing particulate removal. If anyone in the audience wants the details, they'll ask you.

Case Study: Three-Minute Speeches and Their Outlines

To demonstrate the transition from outline to speech, I developed three sample speeches based on the technical presentation, decision-points speech, and persuasion speech that were outlined earlier in the chapter. They are short examples of how those speeches might turn out. It is unlikely that a presentation at a technical conference will last only three minutes, but including a 60-minute speech here is not practical.

Each speech is designed to be three minutes long, so they are a bit short. If you are in a situation where a longer speech is called for, I hope you can see where to expand the speech to meet the time limit. See if you can outline the speech as you read it; you can check your work against the outline that follows each speech. This should help you see how to move from developing the speech structure to delivering the speech.

Example 1: Service Club Presentation (Derived from a Technical Presentation)

Mike, a public works engineer, was asked to give a brief presentation about the water treatment plant upgrade recently completed for the municipality where he works. He was contacted by another engineer, Harold, who is a member of the local Rotary Club and attended Mike's presentation at the statewide engineering conference. He tells Mike that he will only have a short window to speak about the project because of other club activities during the meeting, but he figures a three- to five-minute overview speech would be great. Mike agrees to speak, but it will mean revising the presentation he gave at the technical conference. He updates his objectives and outline to fit into the time frame and renovates his slides to prompt him through the speech. His speech ends up sounding like this:

> Thanks to the Rotary Club for having me today, and thanks to Harold for asking me to come speak to you. As Harold said, my name is Mike Johnson and I am an engineer at the Hometown Public Works Department. I had the pleasure of being the project engineer for the recent water treatment plant upgrade.
>
> We have a good water treatment plant, providing a valuable service to our community. All across the city, whenever people turn on the tap, they get clean water to use for cooking, bathing, and drinking. When there is a fire in our community, the fire department knows that they can hook into the water system and get the water they need to fight that fire. The investment that has been made in the city's water system provides a great payback to the city. However, to ensure that we keep getting the best value for our dollar, we must plan and invest in the water system to keep it up to date with community needs.
>
> The recent plant upgrade project addressed two different problems. First, public works realized the water treatment plant's current capacity could be surpassed in the next five to seven years if the city continues to grow at a 2% per year basis. Second, since the last water treatment plant upgrade done in 1980, major change has been made to the regulations that govern water treatment, and another major change is on the horizon. This rule change is also in the five-to-seven-year planning horizon.

The city selected Apex + Peterson Engineering to design and oversee the construction of the upgrades. Their original cost estimate of $3.4 million proved to be right on target. The money to pay for construction came from a 3.5% increase in the water fee as well as an $850,000 grant from the state. Good Brothers Builders was selected as the contractor for the project, and they began turning dirt on the site last fall.

The city decided they wanted an aggressive time schedule, so the project would be completed before the beginning of lawn watering in the spring. That schedule clashed with the geography of the area around the plant expansion area. While the water plant is located far enough out of town that it does not have to worry about encroachment from residential growth, it is adjacent to a granite rock outcropping. Normal excavation for the size and scope of what was proposed would take four months, but if the project was going to be on schedule, that would need to be cut down to one month.

Apex, Good Brothers, and the city decided to proceed with an innovative rock fracturing technique to meet the time schedule after the equipment vendor demonstrated the technique on-site. The fracturing technique proved invaluable in meeting the time frame the city wanted and did not damage the existing water treatment plant. Good Brothers was excited about learning to use the equipment and believes that they will have an advantage over other contractors when similar projects come up in the future. With excavation out of the way, the rest of the construction of the building was pretty straightforward.

What was going into the building for water treatment also turned out to be an innovative approach. To improve water treatment at the current facility, Apex proposed that the city select something called "magic box technology" to meet current and future regulations. The upside of the magic box technology is that it can meet the current requirements and be retrofitted to meet upcoming standards. However, it was slightly costlier than the alternative "black box" technology. Estimates on the cost of each technology were $1.1 million for the magic box and $950,000 for the black box. The city chose not to go with the black box technology because, even though it is a cheaper alternative, it did not allow for any flexibility into the future.

The plant has been operational for five months, and things are running smoothly after the initial period of learning how to optimize the magic box technology. The facility works great, and the water quality has improved so that we can meet the new regulations. We feel like the project has protected the capacity of our water supply for the foreseeable future and are proud to be using an innovative approach to providing healthy drinking water.

I know that was a brief overview, and if any of you have more questions, I will be happy to visit with you after the meeting or at a later date. Thank you for the opportunity to come speak to you about the water treatment plant.

The speech outline:

- Introduction
- Body
  - Objective 1—Project History
    - How and why this project was undertaken
    - Who funded; who designed; who built
  - Objective 2—Exceptionalism
    - What made the project time line difficult?
    - What makes the engineering design and project process rare?
    - What geographic setting and elements provided major obstacles?
- Conclusion
  - How the project is working since completion

Example 2: Decision Points

The second speech is Vivian's road reconstruction project. She is giving the city council an update on the project progress and asking them to make a decision about whether to include a bike lane as part of the project.

Mayor and members of the council, my name is Vivian Mitchell, and I am a transportation engineer at Blaurnes Engineering. As you may remember from previous meetings, our firm is working on the reconstruction of Pretty Busy Street. We were selected to perform the design and construction management for the road upgrade.

Anyone who has driven on Pretty Busy Street knows the street is failing to move traffic at an acceptable level. Because the street's capacity is not meeting demand, traffic has become slow-moving and dangerous. The intersections on Pretty Busy Street have a level of service that is below the acceptable standards. Engineering analysis determined that upgrades are needed to improve the functionality of the street, and so Our Hometown decided to improve the street.

Most of you are aware that the reconstruction of Pretty Busy Street has been a part of Our Hometown's capital improvement plan for the past three years and that the project has been in motion for about the past six months. The budget for the road reconstruction is $5.2 million. This includes $500,000 for design and survey of the road, $350,000 for construction administration, and $4.35 million for construction cost. With the design of the road 75% complete, the engineering portion of the project is on budget.

Our firm was brought in four months ago to begin the design process that will allow for the road to be reconstructed next summer. My firm began the design process by meeting with public works to discuss the scope of what was to be accomplished. From that meeting, we had our engineers begin the process of designing the upgrades. We had surveyors out to survey the road, and the engineers began running traffic modeling scenarios. We started putting together design alternatives for how the road might look after it was reconstructed. In consultation with public works, we were able to arrive at a road design that met most of the goals for the roadway improvement.

Funding for the road upgrade is expected to be completed with a mix of roadway maintenance fees, use of reserve funds, and a loan from the Department of Commerce. Roadway maintenance fees will cover $3.75 million of the cost. As you are aware, the council voted to allocate $750,000 in reserve money to ensure that the project was able to begin. The balance of the expected money needed, $700,000, will come from a loan program at the Department of Commerce that provides money to cities for infrastructure projects. The interest rate will be 1.5%, and the loan will be paid back over 10 years with money allocated from the roadway maintenance fund.

In consultation with the Public Works Department, we have settled on the exact boundaries of the project. As you can see from the slide, the project will begin on the west side of the intersection with Anderson Street. This will allow us to redesign and construct a new intersection with Pretty Busy Street. That particular intersection sees a higher-than-normal rate of accidents, and the upgrade will increase intersection safety. The project will end with the upgrade of another intersection—with Masters Street. The slide shows all of the businesses that will be affected by the construction.

I assure you, there was much discussion in our office about the disruption to Harlen's Coffee and Doughnuts. We know that many people love to go there and are not looking forward to the road being torn up in front of the shop. However, we will be able to make improvements to the ingress and egress at the shop that should help them get customers through the drive-through more easily and avoid any backups onto the street.

I want to wrap up by covering our next steps and where we need your help. We will be delivering the 95% complete set of plans to public works by the end of the month, with the hope that the project will go to bid in about two months. We are also working with public works to get the loan application to the Department of Commerce completed and submitted. In order to get the final set of plans done, we need your help.

As you know, the Our Hometown Trail Plan calls for a bike lane adjacent to Pretty Busy Road. One of the reasons I am here is to ask whether you would like to see a paved bike and pedestrian

> trail on the east side of Pretty Busy Street. Up on the screen, I have a slide that shows what the road will look like both with and without a bike lane. If the bike lane is included, we will have to widen the road by 15 feet. For most of the road length, the right-of-way already exists to put in the bike lane. However, on the north 500 feet of the road, the city would have to purchase extra right-of-way. We estimate that the extra right-of-way would cost around $15,000 and the bike lane would increase the project cost by about $125,000.
>
> The current estimate for the cost of this project is about $5.2 million, so the cost of the bike lane is not a big part of the overall cost. However, it is an extra cost, and funding for the additional cost of the bike lane will have to be approved by the council.
>
> Thank you for your time this evening. I am here to answer any questions you may have. I have a slide on the screen that shows you options 1 and 2, and I would ask that you choose one of those options so that we can continue to move forward with project design.

Fig. 3-2 shows the outline of Vivian's speech.

Example 3: A Persuasive Bridge

The third speech concerns a proposed bridge replacement. The speaker is with an engineering firm that has been hired to work on the bridge design. However, the engineer also knew she would be delivering speeches on the bridge replacement project because of tepid community support for the replacement. This is a short speech designed to be used in situations where the engineer is addressing an audience that may have concerns about the project.

> Good afternoon. I'm excited about the opportunity to visit with you about the replacement of Old Lonesome Bridge. My name is Brianna Smith, and I work for XYZ Engineering. We are the firm that has been hired to design the bridge replacement, and I want to share with you a few of the reasons the bridge needs to be replaced.

The bridge was installed to provide easier access to the north side of the river, and it has completed its mission well. The north side of town has grown immensely, and a tremendous number of people use the bridge to access housing, businesses, the ballpark, and the RiverWalk corridor. In the late 1990s, an inspection of the bridge discovered it was beginning to show signs of deterioration and discussion began about replacing the bridge.

Old Lonesome Bridge was installed in 1948, just after World War II, and was refurbished in 1974. The bridge was designed to last for 50 years, but the growth on either side of the river prompted the upgrades that were made in '74. We are 30 years beyond those upgrades, which focused on increasing capacity and did not focus on extending design life. With an extra 15 years beyond the bridge life span, it is beginning to show its age. Abutments and decking are beginning to crack, and the piers are being worn down by the water. If this continues, truck traffic will have to be limited in the near future and lane closures for cars will not be far behind.

While no one will be happy with restricted travel on the bridge, a complete shutdown would be even worse. About 35,000 cars per day travel across the bridge; about 20% of that traffic is trucks. If bridge deterioration continues to the point where it is not safe to travel on the bridge, people will be forced to drive eight miles out of their way to use the Interstate Bridge. That bridge is also near capacity, so adding additional traffic will cause delays, adding even more commute time. Most people will see an increase of about 30 minutes on their trip across the river.

It will also mean that emergency vehicles using the bridge will lose valuable time transporting people to and from the hospital and airport. When a person on the north side needs a trip to the emergency room, he or she will lose the important minutes that matter getting to the hospital. A person at the hospital who needs to get out of town via the airport will also lose those precious minutes if he or she cannot cross Old Lonesome Bridge. School buses will have to run earlier in the mornings and later in the afternoons to accommodate the detour.

Because of the problems I have been describing, we have put together a proposal to replace the bridge. The city is proposing to add a new bridge in the same location at a cost of $17.5 million. The new bridge will be wider and safer; it will also provide for bicycle and pedestrian crossing, plus more capacity for vehicle traffic. The project will entail removing the piers from the river, helping to eliminate some of the effect that the bridge has on the river. The only downside is, the bridge will be closed for two months. We will be using a new bridge construction method that allows much of the bridge to be constructed off-site and then installed over the course of the two months. While no one is excited about the bridge being closed, two months is much better than the 10 months that bridge replacement would take using conventional construction methods.

We studied several alternatives before coming forward with the proposed bridge. We could install the bridge just upstream or downstream of the current bridge, but that will require relocating several buildings on both sides of the river and realigning the streets. Adding in the cost of those building replacements and street realignments makes this option cost-prohibitive. We also looked at moving the bridge downstream half a mile, but the river in that location makes bridge construction problematic and also drives up the cost of the project. Moving the bridge upstream becomes problematic because of power lines crossing the river and interference from the railroad bridge.

I will finish with the good news and the bad news. The good news is we will get a new bridge to replace an aging structure that is important to access on both sides of the river. Construction time will be minimized, and walkers and bikers will be able to use the bridge safely. The bad news is the project is not free and the city will be raising the bridge fee to pay for the new construction. During construction, the detours required will not make for a pleasant trip for those using the bridge to get across the river. Lucky for us, we only have to replace Old Lonesome Bridge every 60 years or so.

If we can get people on board with the bridge replacement, we expect to be breaking ground in early spring and have people crossing the bridge in May. Thank you for taking the time to consider all of the important aspects of the Old Lonesome Bridge replacement project. If you have any questions, I would be happy to answer them.

The outline for the speech would look something like this:

- Introduction
  - Introduce self and preview the presentation
  - Role the bridge plays in the transportation system
  - History on how long there has been talk of bridge replacement
- Body
  - Objective 1—Safety
    - Age of bridge and design life of bridge
    - Deficiencies of this bridge
    - How deficiencies can lead to decreased level of service or bridge collapse
  - Objective 2—Resiliency
    - How many cars and trucks travel across bridge daily and/or yearly
    - Other ways to cross the river
    - Is bridge a vital link or nonessential?
    - How long it takes to detour around the bridge
  - Objective 3—Best Alternative
    - Preferred alternative for bridge replacement
    - Other alternatives for bridge replacement
- Conclusion
  - Presentation Summary
  - Reasons the preferred alternative is the best option

The Speaker's Toolbox—Visual Aids

Engineers use a variety of tools to complete their daily engineering tasks. A common list includes a calculator, a computer with the right software, and books of engineering standards. These tools support the practice of engineering and make for a productive engineer. The same principle can be used in public speaking. You can use various tools to support you in making your presentation. The outline template is one such tool. Various types of visual aids are another important tool. Sometimes, a diagram will communicate an idea faster than a verbal description. Sometimes, your audience may grasp an idea more quickly if you draw it for them. And do not forget that good visuals can simply help you keep your audience's attention.

Visual aids come in many forms but really flow from the idea that a picture is worth a thousand words. Instead of describing what a new road may look like, show your audience several pictures, and they will immediately grasp what the road will be like when the construction is complete.

Visual aids include pictures, sketches, detail drawings, artist renditions, a computer presentation, handout sheets, product samples, maps, and videos—just about anything that can be shown to the audience. A 24 × 36-in. stiff foam-backed poster with a schematic sketch may be set on a tripod to show the audience what a new water treatment plant will look like. A computer and projector may display a three-dimensional model of a wastewater treatment plant update. A sheet of paper may be passed out that has several potential typical sections for the design of an urban road. Whichever method you choose, keep in mind that more is often *not* more. Do not overwhelm the audience with a mess of pictures and drawings so that they get lost among the visual aids. Select the visual aids that best convey your message, and use only what is needed. Make sure your visuals are vibrant and uncluttered, can be seen from a distance, clearly delineate what the picture is showing, and have good labels.

Handouts

There are two schools of thought regarding whether to hand out visual aids or written information before a presentation. One school argues that handing out papers before the presentation means the audience will read the handout instead of listening. The other school insists that handouts enhance the presentation, letting people follow along and take notes during the presentation. Decide what works best for your presentation. I do not think there is a hard and fast rule concerning when to pass out written material related to the presentation; rather, it is something to consider based on the type of presentation you are giving. Each situation is different, but there are general guidelines to consider when providing handouts to the audience.

If you are giving a technical presentation and it may be difficult for the audience to see the information, they will appreciate having a handout. The handout may include details or supporting information not included in the slide deck. This allows the audience access to the information while the presentation is not cluttered with references, charts and graphs, case studies, or contact information. If you need or want the audience to refer to the information at a later date, a handout is a good way to provide them a reference point.

If the presentation is less technical and is designed to reach the audience through a more emotional appeal, handouts are probably not warranted. You want them listening to you, not reading a handout while you are speaking.

The type of audience at the presentation provides another guideline for handouts. A public audience with a wide variety of people may not be a great one for detailed, technical handouts. Instead a short summary handout could be more appropriate if handouts are needed at all. A group of engineers wanting to see technical data would be a good audience for handouts. Knowing the audience will help you decide if they are better off with or without a handout.

If you decide handouts are important, when you pass out material is a consideration. If there is material you want the audience to see as you present, give it to them before the presentation. If feasible, have them ready for each person as they enter the room. Keep in mind, audience members who are reading or looking over your handouts are not listening to the presentation. If you are hoping the audience takes some notes during the presentation, a handout is a great way to encourage them to do so.

In rare cases, passing out handouts during the middle of a presentation is warranted. If you want them to listen and then look at the information, you can stop the presentation to pass out handouts. It is good practice to have someone available to assist in passing them out. You can lose some audience attention during this process, so use this method sparingly and have a plan to get their attention after the handouts have been distributed.

If you want to give the audience something to leave with after the presentation, you will need to mention to them that they can grab a handout before they leave the room. Let them know what is in the handout and tell them that it will give them a reference for the future.

If you are passing out handouts, make a good impression by putting some effort into creating a good handout. Develop a handout that serves as a summary of the key points and additional information. It is good practice to avoid handouts that are printouts of the slide deck. They can be unreadable if they are too small, or they will be read by the audience as a way to avoid listening to the presentation.

It is good practice to keep handouts as short as possible, one page if you can. A few key points and contact info should be on one page. If you are handing out additional information not covered in the presentation, extra pages are warranted. A well-designed handout will be a nice complement to a good presentation.

Presentation Software

Audiences have come to expect speakers to use a slide deck as a visual aid during presentations, and few speakers do not use the ubiquitous computer, projector, and slide deck. Most engineers do not have a problem producing visual aids with pictures, sketches, or diagrams, and plenty of other resources can help with the design and execution of those items. However, one visual aid is so commonly used (and misused) that it deserves extra attention here: the slide deck presentation.

Presentation software is available from several different vendors and in order to not recommend one program over another, I will use the generic term "slide deck" to describe the series of slides used in a presentation. (The term derives from how presentations used to be given. Before presentation software, pictures were turned into photographic slides, and those slide images were projected on a screen using a slide projector. A collection of slides was placed in a carousel, or slide tray, and the carousel was placed on top of a projector. The slides in the carousel were called the "slide deck.")

Among the available types of presentation software, the oldest and most commonly used is Microsoft PowerPoint. It is part of their suite of business software products and often comes loaded on a computer bundled with Word, Outlook, and Excel. Other presentation software tools that are regularly used are Google Slides in Google Docs, Prezi, and SlideRocket. Each software has pluses and minuses, so pick the one that works best for you.

Presentation software has several advantages over the precomputer methods:

- The programs are relatively easy to learn and use.
- Software enables a visually pleasing presentation to be designed with minimal effort.
- It is easy to add text and graphics to the presentation.

Although there are drawbacks and pitfalls to using presentation software (more on that below), the advantages are such that almost everyone at times uses the software to make presentation visual aids.

The major drawback in using presentation software is that it becomes easy to avoid the hard work of preparing and delivering a good presentation. Instead, speakers throw some slides together in the software, and that is the beginning and end of their presentation preparation. Too many people seem to think that once they create a slide deck, all they have to do is read from their slides and their presentation will be great.

Yet how many of us have suffered through presentations where the slides were simply read to the audience? About one minute after a presenter starts to read the slides, the audience is already thinking about something else. When it becomes clear that the speaker is going to read the speech, most audience members read the handout or the slide and then tune out of the presentation.

Think of it this way. Most people have the ability to read 250 words per minute, and most people only have the ability to speak at the rate of 125 words per minute. Thus, when a speaker reads slides to an audience, the audience knows instinctively that it is much quicker to read the slides than to listen.

When you speak, it is your job to provide a reason for the audience to listen. You need to give them something from listening that is better than what they would get from reading (otherwise, just give them the handout and go home). If an audience is willing to give you their attention, you must make the effort to offer information they cannot get from reading. Yes, that means your speech should be informative, interesting, educational, and possibly entertaining. Otherwise, everyone is wasting time and effort. Effective speakers strike a mutually beneficial agreement with their audiences—listen to me speak, and I will provide a speech worth listening to.

Do not let the visual aids get in the way of the relationship between you and your audience. Think of presentation software as a tool, not an end in itself. Just because you are using a slide deck, do not assume you are giving a good presentation. It is the same as an engineer designing a bridge using a computer. Just because you are using a computer, the bridge design is not necessarily any better. If you do not know how to design a bridge, it does not matter what computer or software you are using; you will end up with a failing bridge. If you do not know how to give a good presentation without a slide deck, you will not know how to give a good presentation with a slide deck.

If you have completed the process of designing an outline of a speech, you are well on your way to developing great slides. But you should keep the following tricks and hints in mind.

Beware of Defaults

Grabbing the speech outline, you sit down at the computer, open the presentation software, and begin the process of creating a visual aid. You notice that the software has been programmed with default settings, allowing you to construct a set of slides with little difficulty. These default settings provide a font, font size, themed slide background, and bulleted lists to fill out. Usually, there are text boxes that prompt you to place text in those fields. It's easy for you to just fill in the blanks, create enough slides to complete the slide deck, and feel you have constructed a good presentation.

However, the default settings can fail you in a few ways. First, it allows you to write as much as you choose on a slide. You can cram as much text on the slide as you want, and the software will resize the text to ensure that it all fits on the slide. That means one of two things will happen. Either the audience is trying to read the slide rather than listen, or they are struggling to do so because the font is too small. Neither case promotes communication with the audience.

Second, when you create your slide deck from the default settings, you usually end up with a series of slides full of bulleted lists—the same dull visual aid used in every other presentation. Using the default slides helps you prepare an average-looking slide design, but it does not allow you to maximize the visual appeal of the slides. You won't stand out and captivate your audience.

If you are in a pinch putting a presentation together, default templates and fonts at least provide you with a competent slide instead of an ugly one. However, a slide deck full of default colors, graphics, and layout is never going to make for an impressive presentation. How the presentation looks graphically should be a reflection of what you are communicating with the audience, not what the software developer believes is a nice-looking slide.

Finally, I do recommend sticking with the default fonts because they are chosen for ease of readability. If you have a distinctive font you want to use, make sure it is easy for the audience to read from their seats. While you do not want your listeners spending time reading the slides, you do want them to be able to read the words you have on the slide.

Following the default settings is the easy way to use presentation software, but not the most effective way. You do not have to be a design guru to get away from the default settings. Following a few guidelines and making the effort to design a good slide deck will improve your presentation. Visual aid development has to start somewhere, but designing an effective slide deck means moving beyond the default settings.

Beyond Defaults

To begin building a better visual aid, start with the default slide, but do not let the default slide be the only guide to construction. Here are some tips to help you construct an effective slide that is fresh and appealing.

Text. The text you place on the slide should be sparse but readable by the audience. It may look good on a computer, but it will look different projected on a display screen. Make sure that the text is big enough to read from a distance. Each slide should contain no more than 24 words; fewer is better. A benefit of few words on a slide is that people will read them quickly and then get back to listening. Using too many words makes the slides hard to read, so your audience works at reading them instead of listening.

If the audience needs to read a slide, quit speaking, and ask them to read the slide. Step aside for a moment so that the audience can take the time to read the words projected on the screen. Ask if everyone has had a chance to finish reading, and then proceed with the presentation. Never read the slide to the audience. Think of the slide as the way for audience members to begin listening again, if their attention *did* happen to wander.

Fonts. Use sans serif fonts such as Arial. They are easier for people to read when they are projected on a screen. Use only one font type throughout the presentation. To ensure that you are using one font throughout the presentation, use "Select All" in the software, go to the font selection box, and select the font.

Color Scheme or Theme. Choose one and leave it the same throughout the presentation. Selecting a default theme is fine if you have no alternative. Your company may have a theme with its branding, or you may want to develop a color scheme that better suits the presentation. If you want to change the theme, change it throughout the slide deck. The color schemes are developed to look pleasing to an audience.

Choose one that is visually pleasing and does not have colors that conflict with the pictures or text color. Then use it consistently. Keep your slides clean, simple, and neat.

Graphics. Each slide should have some sort of graphic. Graphics include pictures, graphs, charts, or any other type of visual element. Take the time to find the right picture to help your audience understand your message. Graphics should be simple and understandable. If there is too much information on a chart graph, for instance, the audience will have a hard time absorbing the information (assuming that they can see the detail). If you are required to use a complex graphic, spend time explaining the graphic to the audience.

Motion. Presentation software offers a vast array of choices for adding movement and sound to the slides. Avoid using them. Having the text drop in from above (or swirl in or race across from the side) does little to enhance the presentation and can be distracting. These extras offered by the software are seldom needed and do nothing to improve presentation effectiveness.

Quantity. Use only the number of slides needed to communicate your message. A good guideline is no more than one slide per minute of the presentation. If you are doing a 20-minute presentation, use no more than 20 slides. If you are doing a 20-minute presentation and plan to talk about each slide for five minutes, then you need four slides. Each presentation is different, so you need to think carefully and then practice. It will become clear if you have too many or too few slides for the time allotted.

What an Effective Slide Looks Like

The most effective slides developed for a presentation combine a few words and a graphic or two to reinforce those words. At the top of the slide should be a title or header, limited to one sentence, or two at the maximum. This should be the main point reinforced by the slide graphic. The type size should be big enough for the audience to read. *The fewer words, the better.* Below the line of text should be a graphic that supports the slide's main topic. If you do not have a graphic that corresponds with that portion of the presentation, try to be a little creative in coming up with a graphic that works. If you really cannot find a graphic that works, consider a blank slide or turning the few words into the graphic. It is easy to insert a blank slide into the deck, and most presentation software has shortcut keys that allow you to turn the screen black or white with a push of a button. (In PowerPoint, pushing the "B" key on the keyboard puts up a black screen and pushing the "W" key puts up a white screen. Pushing either key again, or the Esc key, returns to the slide that was up when you pushed the key.)

Below are a few example slides that show two approaches to slides: all words versus words plus pictures.

Fig. 3-3 shows two slides describing an elephant. One slide gives several bullet points and because we all know what an elephant looks like we can picture that in our heads. The other slide gives a picture of an elephant with a couple of maps showing where elephants live. This slide is much more attention-getting. It allows you to talk about your bullet points rather than be tempted to read them. And it encourages the audience to see what you want them to understand—what an elephant looks like and where it lives.

Fig. 3-4 shows two slides for a presentation discussing a road cross section. If you are presenting to an audience of transportation engineers,

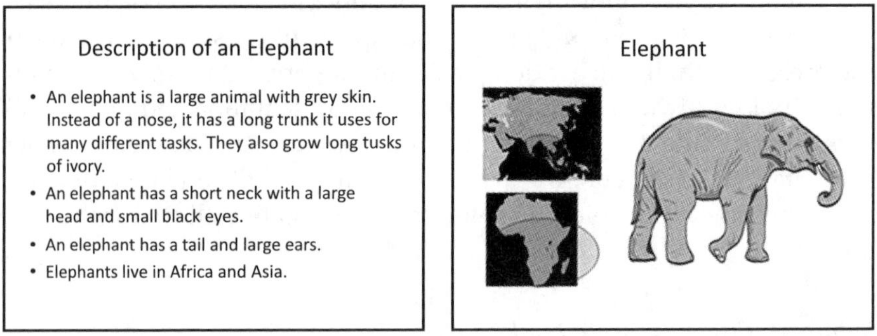

Description of an Elephant
• An elephant is a large animal with grey skin. Instead of a nose, it has a long trunk it uses for many different tasks. They also grow long tusks of ivory.
• An elephant has a short neck with a large head and small black eyes.
• An elephant has a tail and large ears.
• Elephants live in Africa and Asia.
Elephant

Fig. 3-3. Elephant Slides

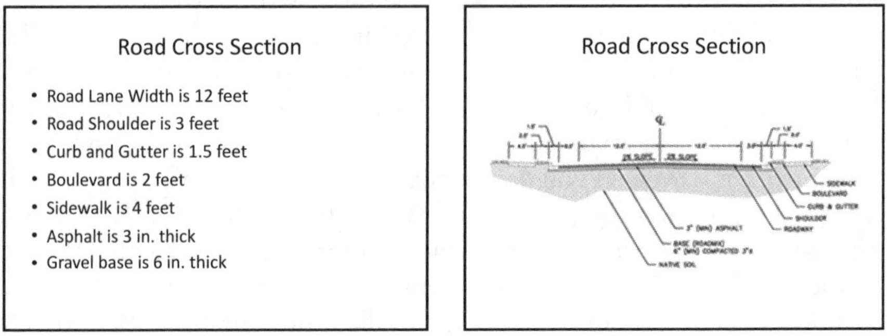

Road Cross Section
• Road Lane Width is 12 feet
• Road Shoulder is 3 feet
• Curb and Gutter is 1.5 feet
• Boulevard is 2 feet
• Sidewalk is 4 feet
• Asphalt is 3 in. thick
• Gravel base is 6 in. thick
Road Cross Section
SIDEWALK
BOULEVARD
CURB & GUTTER
SHOULDER
ROADWAY
NATIVE SOIL

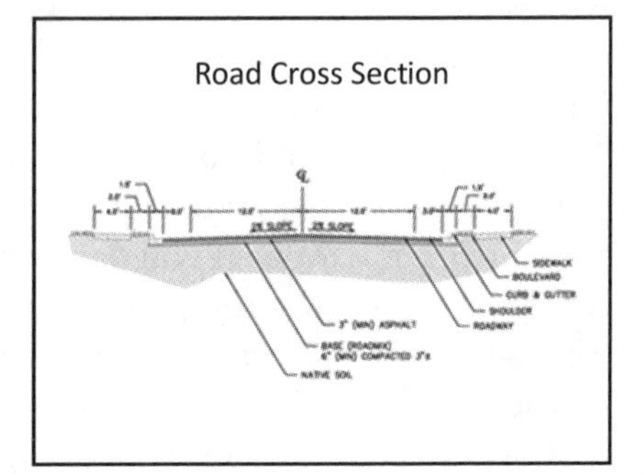

Fig. 3-4. Road Cross Section Slides

Water Tank

- Water Tank will be 35 feet high with a diameter of 30 feet.
- Tank will be constructed of steel
- Tank will sit above ground and be located near Elm Park
- Tank will be meet AWWA and NSF requirements

Fig. 3-5. Water Tank Slides

Project Location

- Project is located 2 blocks north of Grand Ave along Avenue D and Avenue C
- Project is bounded by 6th Street West on the east and 8th Street West on the west
- Project is located one block from High School and Park

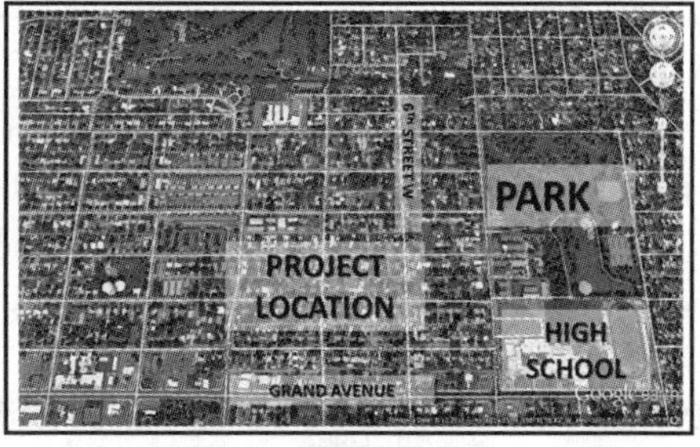

Fig. 3-6. Comparison of Bullet Points and Photographs to Describe a Project Location

then the slide with the bullet points will likely get the job done. However, if you are presenting to a public audience, then the slide with the drawing of the cross section will be much more effective. This audience is not able to paint a picture of a road cross section in their minds, as they would with the elephant. The picture promotes understanding and helps the speaker communicate.

The final four slides (Figs. 3-5 and 3-6) are examples of bulleted lists versus pictures for a water tank and project location. Notice that if you were using the graphic slides, you could use the bullet points for prompting while speaking.

Case Study: A Slow, Painful Death by PowerPoint, aka the Water and Wastewater Master Plan Update

During my term as a council member, the city hired an engineering consultant to develop a master plan for the water and wastewater systems. The city tends to use the same engineer each time they upgrade the water or wastewater plants or, in this case, plan for future infrastructure. They hire the same firm because it has a great engineer who provides terrific service. He is a smart and talented engineer and has always delivered excellent projects.

After the firm completed the water and wastewater master plan, a presentation on the plan was given to the city council. The presentation's objective was to report on what was accomplished and describe where the city is headed concerning the water and wastewater systems. The presentation took place at a work session because it was a less formal setting and a better place to have a discussion. The evening the water and wastewater master plan was on the agenda, I was looking forward to hearing about the completed project and where the systems were headed.

The engineer did the right thing and had his slide presentation set up before the meeting started. There were two other short items on the agenda before he spoke. I turned my chair to face where he was speaking and was ready to learn about the future of our water and wastewater infrastructure. It was a good thing that I like water and wastewater infrastructure issues. Although the speaker was a good engineer, he was not a good public speaker. I stayed mostly engaged because I am interested in the topic, but the other council members were not as interested and did not listen.

When I looked at my colleagues, I saw a disengaged group. Ten out of eleven council members stopped listening to the presentation within three minutes of the beginning of the speech. And then the talk went on and on and on, for at least 30 minutes. If the engineer had been able to judge his audience, he would have known to finish speaking several minutes earlier. The public works director finally put an end to the presentation because the council was about to revolt.

The first problem with the presentation was the engineer's monotone voice. You could have cut to a room full of tax accountants discussing changes to IRS policy on annuity exchanges and found a more dynamic speaker. The engineer spoke with his back to the council and read the PowerPoint slides with nary a change in inflection. All of the slides were bullet-point lists full of engineering details that were not designed for this audience. The second problem: The presentation needed to be much more dynamic, highlighting the report's important sections and grabbing the attention of council members. It could have used stories and graphics highlighting the report's content. The audience needed to learn something from the presentation, but the level of detail was not appropriate for them.

As for the presentation content, the engineer opened with the methodology used to develop the report, continued with an analysis of engineering concepts, and finished with several unreadable charts and graphs. A better approach would have been to start by discussing the city's future and how it will be affected by investment in water and wastewater infrastructure. There could have been a discussion about what happens if this plan is not followed, what problems would arise if there are large deviations from the plan, how growth would be limited, and what kind of funding would be needed to implement the plan. None of these issues were mentioned in the presentation. This audience wanted to know what the report meant for their city and their constituents. Because he did not understand his audience, the engineer failed to cover the topics they were interested in and, ultimately, did not communicate with them. It was a missed opportunity to engage and educate the audience.

---

* * *

Vivian is hard at work developing her slides when Tom walks into her office. She shows him the slides, and he agrees that they look great. Tom can see that she has her talking-point cards sitting on her desk. He asks her what she decided to use as talking points, and they discuss them. He remarks that he can see how her objectives made it into the speech through the outline process. Tom tells Vivian that she has a well-designed presentation and she will be glad she put in the work. He mentions that he would be happy to invite a few people from the office to listen to her give the speech over lunch at the end of the week. Vivian agrees that might be a good idea because she is starting to get nervous about speaking. The best way to combat that anxiety is for her to get as much practice as she can before delivering her speech.

Vivian's well-designed speech will make it easier for her to communicate with the audience and aid with the delivery. Her speech design is an important building block for the delivery of the speech, but she will need to spend time developing her delivery if she wants to convey the information effectively.

Chapter 4

Delivery

Be sincere; be brief; be seated.

—Franklin D. Roosevelt, on speechmaking

This chapter on delivery will cover the ways you can assess your public speaking and provide insight into the skills, strategies, and habits that will help you become an effective speaker. Tom and Vivian are back to provide a look at how Vivian incorporates these elements into her speech.

- You'll learn how to assess your public speaking skills.
- You'll discover the skills that are critical to master on the way to becoming a good public speaker.
- You'll find several strategies you can use during a presentation to make it stand out to the audience.
- You will uncover the habits of great public speakers.

Nothing in this chapter is magic, but you are making a big transition by crossing over from an internal process to an external process. Chapters 2 and 3 covered speech preparation tasks that can be done without having to face an audience. In this chapter, you come face to face with the prospect of being in front of an audience. The fact that an audience is going to watch you is what gives speakers anxiety. How speakers combat that fear has everything to do with mastering public speaking skills and using sound strategies through practice.

Before you can begin improving your skills, you have to know your current public speaking ability. A few assessment tools can be used to figure out where you need to improve. Once you know what you do well and what you need to work on, you can pick out the skills and strategies you should practice and use while speaking.

The three case studies in this chapter provide awareness about dealing with speaking anxiety, the attributes others look for in an engineer, and an opportunity for a couple of engineers to improve on their public speaking and promote the profession. One of them describes an experience that allowed me to become a more comfortable public speaker and provides an interesting take on how anxiety can be useful to speakers. The second case study examines the important role communication skills play during interaction with clients. The final case study illustrates how engineers often drop the ball when they have an opportunity to make a good impression for engineering.

But first, let's check in with Vivian and see how her speech is coming along.

Getting from Design to Delivery

Vivian is putting the finishing touches on the outline and visual aids for her speech when she looks at the clock. The afternoon is nearly over! Empowered by completing the outline, she calls Tom to ask if he would go over her speech outline before they both leave for the day.

Tom peruses the outline and visual aids and says, "Good job, Vivian. I really think the outline is on the money."

She says, "Thank you."

Tom adds, "Did you get help with the visual aids?"

"I had Anna help me. She does such a good job with that stuff," Vivian replies.

"Well you did a great job organizing your speech. Now comes the hard part. Or at least the part I don't enjoy."

Already sensing the next task, Vivian's apprehension shows on her face. Tom sees this and continues, "If you are like me or most engineers, I'm guessing that speaking in front of a group is not your idea of a good time."

Vivian nods. Tom again praises her work in building her outline, adding that her next challenge, delivering the speech, will be easier because of her preparation.

Tom leans back in his chair and tells Vivian about the fear he has seen in the eyes of many of the people who testify before city council. He says, "It's tough to get up there and speak to a bunch of elected officials, many of whom you've never seen before. It takes a special set of skills, strategies, and habits to be good at public speaking, and they do not usually come naturally to the personality type that makes a good engineer."

Tom sits forward and says, "What type of truck was used to deliver the beams on your last bridge job? And what type of crane was used to install those beams?"

Vivian is surprised by the question, but answers promptly. "The beams were delivered by tractor-trailers using remote-controlled steerable dollies because of the bridge beam's long span. The crane was an ABC Model 112, capable of handling the 5-foot height and 97-ton bridge beams."

Tom nods approvingly. "Right. We needed those steerable dollies and the ABC Model 112 crane. They were the right tools for the job. What I want you to recognize is that for this job, this presentation, *you* are the instrument that will be used to deliver the speech. Just as we pick dollies and cranes because they have the right attributes for the job, you need to have the traits that make you an effective instrument for delivering a speech."

Vivian thinks about that for a moment and says, "I suppose that is true, but I'm not exactly sure what you mean."

"You are going to have to work on your speaking voice and make sure it is ready for a public presentation," says Tom. "You should also work on adding elements to the speech that will keep the audience engaged. Finally, and this may give you the most heartburn, you are going to have to practice your speech a few times before the meeting."

"Yeah, I figured that was true, but hoped it wasn't," replies Vivian.

"I can see your hesitation, so let's start with something easier. How about we line up an assessment of your speaking so you can figure out what you want to practice?" Tom says. "I can recommend a couple of people who are pretty helpful and will ease you into getting your voice ready."

"That sounds like it could work. Thanks for the help," says Vivian.

Tom has been an engineer long enough to know that to become proficient at public speaking, speakers need to cultivate the skills, strategies, and habits of good presentations. He is also aware that they can be difficult for his engineers to master. So he encourages Vivian to make a commitment to improving her speaking skills, strategies, and habits. To help her figure out what skills she needs to improve, Tom assigns her two exercises: a self-assessment followed by feedback from a couple of her friends and the folks he knows will be helpful.

Assessing Your Speaking Skills

Tom's advice to Vivian is good for everyone. The self-assessment is a good starting point for identifying which speaking skills you need to improve. But be forewarned: it is hard to be an accurate judge of your own skills and performance. That's why it is so important to get feedback from someone who has watched you practice or (better) give a speech. It is much easier for someone else to observe your performance, evaluate your skills, and provide constructive feedback, outlining the skills that you need to practice or learn.

Self-Assessment

To start the self-assessment process, think about your past experiences in making public presentations. What went well? What did not? Were you able to improve each time you spoke, or do you feel like you always get stuck on the same issue? With honest answers, you can develop a plan for upgrading your next speech. Use the plan as a baseline to build a program for enriching your speaking.

To organize your self-assessment, use the following lists to evaluate key skill areas. I have included suggestions to make it easier to assess yourself.

- What does my voice sound like when I speak? *(The best way to assess this is to listen to a recording of yourself speaking.)*
    - What does my voice sound like? Is it soothing? Irritating? Stimulating?
    - Do I speak loudly enough for people to hear me?
    - Do I speak fluidly or haltingly, with a lot of *ums* and *ers* and *you knows*?
    - Do I speak too quickly or too slowly?
    - Do I speak clearly, or am I prone to mumbling?
    - Do I manage my breathing so that my voice stays strong?
    - Do I speak from my diaphragm so that my voice projects?
- How do my body language and posture appear? *(Try giving a presentation to a mirror or videotaping a practice session. Play the tape back with the volume on and then with it off.)*
    - Do I walk while speaking or stand in one place?
    - Do I rock from side to side or forward and back?

  - Do I use hand gestures? Are they distracting, or do they support what I'm saying?
  - Am I slumping and looking like I want to slink out of the room?
  - Am I upright with good posture, projecting both my voice and confidence?
  - What is my facial expression? Do I look out at my audience often, sometimes, or never?
- What is my state of mind when I give a presentation? *(You probably already know the answer, but not the solution.)*
  - Do I feel panicky? Is my mind a blank?
  - Am I anxious and self-conscious, worried about what the audience thinks?
  - Am I calm and confident, having done my prep work and practiced a lot?
  - Am I resigned, just wanting to get it over with as soon as possible?

One priceless opportunity for self-assessment is available if you are videotaped testifying in front of the local government. Many of those meetings are shown on local access television. Record the meeting yourself and watch it later to see how you did. You can follow the method in the next section, for feedback from others. Fig. 4-1 provides an evaluation form you can use to complete a speaking assessment.

Feedback from Others

The second part of the assessment process is having someone else do an evaluation of your speaking skills. Receiving feedback is always one of the hardest parts of improving anything, including public speaking skills. There are a couple of reasons for that. Feedback can easily feel like negative criticism, and nobody likes that, especially about something in which you may not have a lot of self-confidence. It is uncomfortable to have others point out your shortcomings, even when you know they are trying to help. However, constructive criticism is valuable, and feedback from others is absolutely necessary to help you see what you may not be able to see in yourself. That is the only sure way to improve.

As hard as it is to receive feedback on your public speaking skills, it can be even harder to find someone willing to give it. It takes time for an evaluator to listen to you practice, so you should respect their efforts on your behalf. Make it easy for someone to give you feedback, and be

Presentation Evaluation Form

Presentation	____________________
Reviewer	____________________
Date	____________________
Rating system	1-2-3 = Fair 4-5-6 = Good 7-8-9 = Very Good 10 = Excellent
__________	Start time
__________	Did the speaker make eye contact with the audience?
__________	What was the overall physical speaking voice like—tone and pitch?
__________	Was the voice loud enough?
__________	Was the voice fluid (vs. halting)? Were there few "ums" and "ers"?
__________	What was the pace of the presentation?
__________	Was the speaker moving around the room?
__________	Was the speaker gesturing toward the audience?
__________	Was the speaker using facial expressions?
__________	Did the speaker exhibit good posture?
__________	What was the overall look of the speaker? Did he or she look nervous or anxious?
__________	What was the overall appearance (clothes, personal hygiene) of the speaker?
__________	Did the speaker use phrases and terms I understood?
__________	Did the speaker engage the audience?
__________	Were the visual aids readable and understandable?
__________	What was the overall structure of the speech?
__________	What was the overall rating of this presentation?
__________	End time
__________	Overall length (minutes)
Comments	What did the speaker do well today?
	What does the speaker need to work on?

Fig. 4-1. Presentation Evaluation Form

willing to accommodate the evaluator's schedule. Flattery can go a long way in persuading someone to assess you. Try telling a colleague that you admired the way they made a particular presentation, then ask the colleague to listen to one of your practice sessions and give you feedback. Thank your evaluators for their time, and show your appreciation for their help.

Some people are reluctant to give feedback because it makes *them* uncomfortable. It helps a lot if you can give them an evaluation form or checklist, similar to the one in Fig. 4-1. The evaluation form prompts your evaluator to look at all aspects of your speech; it also provides a neutral arena to discuss your speaking skills. The form in Fig. 4-1 is designed to be used by an audience member to complete a presentation assessment. The form can be changed slightly and used as a self-assessment tool. The form can also be tailored to assess any presentation skills on which you may be looking for feedback. Adjust the assessment form slightly and give it to audience members as you practice a presentation.

In Vivian's case, she should use Tom as a resource to assess her skills and to suggest other people who could give her feedback. By using the evaluation form, Vivian is letting her evaluators know what they should be appraising and encouraging them to provide constructive feedback about the strengths and weaknesses in her public speaking skills. She should schedule a time that is convenient for them to meet and send a thank you note after the evaluation is complete.

Learning by Observing

Another great way to improve your public speaking is by studying effective speeches given by other people. Often the people who are graceful public speakers are paid to speak to the public and have many opportunities to develop good public speaking habits. They are usually effective at presenting their case to an audience. Engineers with good public speaking skills have often developed them through years of practice. They have testified at numerous public meetings, had many discussions with nonengineers away from public meetings, and developed a good sense of how an audience will react to different ideas. Often, these engineers learned those skills to enhance their business and keep clients happy. They have spent hours practicing their skills. Watching and evaluating them will provide insight to becoming a better speaker. You can even try filling out the assessment form as you watch these excellent speakers in action.

Want to take your observations a step further? Watch some of the great presentations on the TED website (www.ted.com). TED—short for Technology, Entertainment, Design—is a nonprofit that sponsors conferences and opportunities for speakers to have 18 minutes in which to give the greatest presentations of their lives. The TED motto is Ideas

Worth Spreading. The conference records all the speeches and posts them on the TED website.

By visiting the TED website, you can browse hundreds of talks about all sorts of topics. View some of the most popular talks with an eye toward evaluating what the speakers did well. How did they begin and end their talks? Did they tell stories throughout their speeches? Did they engage the audience? How little did they use PowerPoint slides? Challenge yourself to watch at least one TED talk every day while you are practicing your public speaking skills.

Yet another benefit from time on the TED website is to discover affiliated events all around the world. These are called TEDx events, and there is a good chance one is located near you. TEDx events are a great place to watch several speeches in a day. Consider attending an event for the day and watching local speakers deliver their speeches. Take an evaluation sheet with you and evaluate each speech. Better yet, if you have been practicing your speaking skills, put your public speaking skills to the test. Take the plunge and apply to talk. If you are chosen to speak, it will be a great payoff for all your hard work.

* * *

With all the planning Vivian has been doing for her speech, she is actually looking forward to getting an assessment of her presentation. She needs to start practicing her speech, and her first attempt will be in the office conference room without anyone else there. She grabs her computer to hook up to the conference room screen, her outline, her note cards, and her assessment form (Fig. 4-1) and heads down the hall. After the computer is set up and the slide deck is showing in the screen, she sets her phone on the table with the audio recorder and timer running and begins speaking to an empty conference room. As she works through her presentation and clicks through the slides, Vivian finds a couple of places where things did not go how she thought they might.

She finishes up her first practice run and stops the timer and audio recording. She checks the time and sees that the speech ran for eight minutes. Her goal was to come in at about six and a half minutes, so she knows some time needs to be trimmed. She takes out the assessment sheet and begins listening to the audio recording. One of the things she notices is that it does not sound like her voice changes much during the speech. She pauses the recording and writes herself a note to work on varying her tone during the speech. After she finishes listening to the speech, she realizes she did not ask the audience if they had questions during the speech. She had planned to incorporate that into the presentation, so she writes on the assessment form that she needs to make that change.

Vivian is satisfied that the first practice of her speech went well. She is apprehensive about the second speech practice because she invited two of her colleagues and one of the people Tom recommended to listen to her. They will be in the same conference room tomorrow at noon to listen to her presentation.

Skills, Strategies, and Habits

Almost all of the aspects I have covered on public speaking to this point have been internal processes. Planning and designing your speech are things that are not done in front of an audience and are largely an internal process. Assessment of speaking skills bridges the gap between an inside process (learning from the assessment) and an external process (having others assess your speaking skills). The next step is to move into the full-blown external process of speaking in front of an audience. After you have planned and designed the speech, you get up in front of an audience and give the speech. To make the transition from an internal process to an external process, it is important to develop the skills, strategies, and habits that support good public speaking. These are the tools you can use to improve your speaking when you are standing in front of a group of people.

The first step is developing the public speaking skills that are commonly used by speakers. Skills are learned behaviors that allow you to accomplish certain tasks. You develop and become proficient with speaking skill behaviors through learning and practice. Each of the skills that I discuss are things that most speakers have to learn—they are not innate behaviors. As long as you work on developing your skills, you will learn the qualities that make you able to deploy those skills. Learning a skill is going to take time and practice, but just like using a computer program, cooking, or driving a car, if you work at it, your skills will improve.

The next step is to use the strategies of a good public speaker. A strategy is a course of action used to ensure that you connect with the audience with credibility and understanding. Strategies are different from skills. Baseball players improve skills like hitting, throwing, and catching, but they use strategies like throwing a curveball when the batter is not expecting it to improve their game play.

Finally, when you have improved your skills and implemented your strategies, you need to use the habits of a good public speaker. Once you learn a habit, you do it almost automatically. It keeps the skills you have learned active and the strategies you use sharp. Habits make skills and strategies stick.

Speaking Skills

Developing a stable of public speaking skills provides a great challenge and an opportunity to advance your speaking capability. Working on a skill set takes time and focused practice, which I will cover later, but if you put in the effort, you will see the results. The skills that I am going to cover will allow you to improve your outward speaking presence and improve your communication with the audience:

- Eye contact,
- Voice control,
- Body language, and
- Dealing with anxiety.

Each of these skills is used to establish good public speaking. Improving communication with the audience means spending time improving them.

Eye Contact

It is a bit surprising how many of the skills and strategies in this chapter are about nonverbal communication. The nonverbal aspects of speaking are important for building trust and credibility with the audience so that they will listen when you are communicating with them orally. One of the most difficult aspects of nonverbal communication is making eye contact with the audience. Human eye contact tends to be an intimate act, so making eye contact with a group of people you do not know can feel especially awkward. The only way to get beyond the awkwardness is to learn the skill of eye contact.

Eye contact is important because of

- Focus,
- Credibility,
- Movement, and
- Feedback.

One of the great benefits of learning to make eye contact with the audience is that it helps you focus on the presentation. It is much easier to concentrate while you are focusing on a subject like an audience member. You will stay on task and avoid letting your mind wander. When you avoid eye contact with the audience, your eyes tend to take in random images not related to what you are speaking about. Your brain tries to process those images, and before you know it, you have lost your train of thought.

The audience is there to see you and to hear what you have to say. They will find you less confident, convincing, and authentic if you avoid eye contact. Luckily, if you focus on them, they will keep your presentation on track.

Good eye contact with your audience is important to establishing trust with the audience. They instinctively know the importance of eye contact because it signals that the speaker is paying attention to them. By showing them you are paying attention through eye contact, they know you think this is important, which creates trust. They want to be included in what you are talking about, and looking them in the eye lets them know you are welcoming them to listen.

Speakers who avoid eye contact with an audience create barriers to communication. These speakers are signaling they do not believe what they are saying and that the message lacks trustworthiness or credibility. Lack of eye contact can really put a damper on a presentation and allow the audience to zone out of a less-than-stellar presentation.

Just before you begin your presentation, find an audience member you feel comfortable making eye contact with to start the speech. Even practiced speakers have a predisposition to let their eyes dart around the room during the first portions of their talk. You can avoid this by picking out the first, or even the first few, people you want to maintain eye contact with as you begin the speech. Look the person in the eye and begin your presentation. After you have looked at them for three to five seconds, move on to the second person you want to have eye contact with.

If you are able to move around the room, move toward the next person with whom you are making eye contact. This will naturally force you to move around the room and not be a stationary figure stuck in one spot. Try to be as random as possible with eye contact, avoiding patterns of left then right then left then right.

Engaging with eye contact will help you avoid turning to look at the projected slides instead of the audience. It will keep you from reading the slides to the audience because you cannot read if you are looking at the audience. Having a few notes in front of you means not looking at the slides and makes for an easy break or prompt to move from looking at one audience member to the next audience member.

Another beneficial attribute of eye contact is assessing the audience feedback. You can determine whether they are listening and interested, or bored and indifferent. A good speaker is able to use this feedback to modify the presentation or to ask the audience questions to keep them engaged. At the very least, you can use the feedback you receive from eye contact with the audience to assess the speaking skills you need to work on.

Voice Control

Although the first thing an audience notices about you is your appearance, the next thing they perceive is your speaking voice. Each and every day, you converse without giving a second thought about how your voice is generated. You do not stop to think about how you are using your lungs, voice box, and tongue to create words to communicate conversationally. Your voice is generated through a complex biological system that allows you to communicate messages, concepts, and emotions. You will use the same instrument to communicate with an audience, but adjustments need to be made between the everyday voice and the public speaking voice. You will change how you use your lungs, voice box, and tongue during a public presentation, so give some thought to how you sound when you speak to an audience and how you can improve your voice. If you want to know what you sound like when you speak, record yourself and play it back. You most likely sound different on tape than in your head, so take the time to listen to the sound of your voice. Here are some of the things to listen for:

- Tone and pitch,
- Volume,
- Speed,
- Breathing, and
- Readiness (that is, a warmed-up voice).

All of the complex movements that create your voice have several aspects to address when you work on your voice-control skills. One attribute is not more important than another, and it is important to work on each aspect. Luckily, there is a great way to assess each of these aspects: Listen to yourself speak.

Tone and Pitch

Whether it is listening to an audio recording of your presentation or watching a video recording, you can easily evaluate your speaking voice. The first thing you want to listen to is the tone of your voice. Voice tone is the characteristic style with which you speak. Humans prefer a voice that changes tones to signal importance, nuance, feeling, and several other subtle communication cues. A voice that does not change tone, a monotone voice, does not provide those cues to communication.

A good presentation provides a change in tone over the course of a speech. You want to hear your voice rising and falling as different points are made. A changing tone tells the audience what is certain or uncertain, urgent or marginal, important or minor. A speaker's tone draws attention to key information and differentiates it from the rest. If your voice sounds enthusiastic, your listeners are more likely to stay tuned in to your message and pick up key parts of the presentation.

Linked closely with tone is voice pitch. Everyone's voice has a range from high to low, and setting your voice in the correct pitch range ensures that your voice sounds genuine. Pitch is similar to the keys on the piano, with a low, deep pitch on one end and a high pitch on the other. Match the pitch of your voice with the tone you are trying to convey. An audience will give you credibility if the pitch of your voice matches what you are saying. Delivering serious information in a low voice matches the audience expectations, whereas delivering optimistic information in that same low voice does not match expectations. Ensure that your pitch matches your tone.

You can use the outline of your presentation to signal where you want your tone of voice to change. If you have one objective you want to stress above all others, change the tone while you are speaking on that subject. Find a way to make your voice reflect the importance of the topic by dropping your voice or making it deeper. If you are discussing a safety issue, use your voice to stress the seriousness of the situation.

If you are wondering how to make your voice do these things, begin by determining what you want to impress upon the audience about each objective. If you want to stress the importance of one objective over another, then your voice needs to contain urgency when you speak about that objective. If you want to stress how certain you are about an objective, then your voice needs to sound certain. Your voice will sound different if you are speaking with urgency or certainty.

Let's take Vivian's speech as an example. She begins with an introduction in which the tone of her voice will be pleasant and welcoming. She is using her tone of voice to warmly connect to the audience. As she moves on to Objective 1 (budget, time line, and funding) her tone should reflect certainty about these items. If things are not going according to plan, she should add some concern to her tone as she explains why things are off track. She should switch the tone back to certainty when she talks about how they are going to fix the problem. Objective 2 (project limits and businesses disrupted) should be expressed with sympathy for the people who will be disrupted during construction. Vivian needs to use an inquisitive tone as she talks to the council about the project next steps. She needs to signal to them that she is looking for answers to the questions that remain about the project. Finally, she needs to finish the speech with a tone of voice that conveys competence. The changing tone of her voice will help keep her audience engaged.

They will not be engaged if her speech sounds indifferent or monotonous. A low-energy speaker, no matter that they have the greatest engineering design known to humankind, will be described as flat, uninteresting, and limited. If you, or others, describe your speaking voice as monotonous, then you need to work on changing the tone of your voice.

Overcoming a monotone speaking style will require learning how to change the tone of your voice to emphasize different presentation aspects. Everyone can speak in a low voice and a high voice, but ensuring that you do not speak in a monotone voice means moving up and down the scale of voice tones. People who do not speak in a monotone can do this without *thinking* about changing the pitch of their voices. If you are a monotone speaker, though, then spend time thinking about changing the pitch of your voice while speaking. After some practice, you should be able to modulate the tone of your voice without having to think about it.

The easiest way to change your monotone speaking voice is to identify the attributes you want to add into your speech. To practice speaking with variety, try reading poems or children's books out loud. Find your favorite Dr. Seuss book and read it aloud. Follow the cues in the book as to what you are supposed to be expressing, and add them to your voice. When the characters whisper, your voice should be quiet. When they are scared, your voice should sound like a person who is scared. Record the reading of the book and listen for excitement, suspense, drudgery, or any of the attributes you are trying to convey to make the story interesting.

Volume

You should also be assessing the volume of your voice, paying attention to the loudness or softness required by the space in which you are speaking. How loud or soft you speak will vary depending on the room and the audience. As you enter the room, judge how loud you are going to speak so that someone sitting at the back of the room can hear you. A presentation to a small conference room of people with the doors closed does not require the same volume as a speech given in an acoustically poor room to a group of senior citizens. If you are unsure about the volume you need, ask the audience if they can hear you. If they cannot, you will not communicate with them.

You should always try to project your voice to the back of the room by pretending you are trying to hit the back wall of the room with the sound coming from your voice. If there are microphones in the room, speak into the microphone and ask if everyone can hear you. As long as you are close enough to the microphone to amplify your voice, you do not need to be as loud. Let the microphone do the work of carrying your voice to the back of the room. You should consider practicing with the microphone, if possible, prior to speaking to the audience, especially if you are going to be wearing a portable microphone.

Speed

A close companion of volume is the speed at which you speak. A nervous and unprepared speaker will often rush through her speech to get it over quickly. You should be able to hear if you are speaking too fast by listening to or watching a tape of the speech, or by getting feedback from the audience. Slow down if you hear yourself speaking too quickly or if you pick up cues from the audience. Keep track of the time it takes to give the speech each time you practice the speech, then compare those times. If you are going too fast, slow down and make sure the speech is longer the next time you practice.

It is rare for a speaker to go too slowly. However, if you ask others to assess the speech's pace and they say you need to speed things up, then keep track of the time it takes to give the speech and speed it up each time.

Breathing

The bellows that fuel the fire of the speaking voice are the lungs. The lungs force air through your speaking system to make it audible. How you breathe during a speech is an important part of your voice because breathing allows for vocal variety. Speaking properly requires that you breathe deeply. Ensuring that you breathe with your diaphragm as relaxed as possible allows your lungs to fully expand and creates the reservoir of air you need to control your speaking voice.

One of the benefits of breathing properly is that deep breathing reduces tension and provides focus. Good breathing will keep you from speaking too fast because you will be taking short pauses (about 0.5 seconds) to inhale. Start speaking after you have taken a deep breath and, as you practice the speech, make notes of where you need to take a quick pause for a good inhale during the presentation. If you practice your breathing, it will become second nature and improve your speaking voice.

Readiness

Speaking tone, volume, breath, and speed are the vocal instruments of your presentation. Having an effective voice includes making sure the vocal chords are warmed up and ready to go before speaking. Before going on stage each night, good actors warm up their voices backstage before performances. That way when they enter the stage, they are ready to speak clearly. The same goes for giving a speech: ensure that your voice is warmed up before getting started.

Several exercises can be used to warm up the vocal chords.

Place your hand on the top of your head, keep your mouth closed, and create a HMMMM sound with your voice box. Placing a hand on the top of your head ensures you are doing the exercise correctly because you should feel the vibrations on the top of your head.

After a couple of minutes of voice box vibrations, try repeating the phrase "Red Leather, Yellow Leather" over and over for about a minute. With this exercise, the lungs, tongue, and lips are getting warmed up. Finally, try a short run through of the speech's opening section, listening to the tone of your voice as you speak the lines. Go through the speech's first 10 lines listening to your voice and how it is performing. If you do not like what you hear, repeat the lines, paying particular attention to fixing what you did not like.

Taking the time to warm up your voice has a double benefit. First, when you step up to give the speech, your voice will not falter or crack. You can step up to the podium with a clear voice and make a good impression. Second, it will give you something to concentrate on before the speech, which should help you relax. A good vocal warm-up will give you a chance to not only place aside any anxiety, but also help prepare the most useful speaking tool in your toolbox.

Harnessing the complexities of the human voice fosters improved communication, but you have to practice the skill of voice control if you are going to improve as a speaker.

Body Language

From the moment you step in front of the audience until their attention wanders, the audience will be reading your body language to supplement what they are hearing. Communicating with an audience involves both oral sounds and body language. To be a good speaker, you must engage the audience with your body, face, and eyes during the speech, standing up straight and projecting your voice into the last row.

I have already discussed the importance of making eye contact with audience members, and eye contact goes hand in hand with the gestures and facial expressions the audience sees during important points. This allows the audience to connect with you, find you credible, and become active listeners. The body-language skills important to a public speaker are

- Posture,
- Gestures, and
- Facial expression.

Posture

Good body language begins with good posture. Standing properly helps you in many ways. It allows you to fill your lungs with air to breathe properly, it points your eyes toward the audience, it makes it easier for you to move around, and it gives the perception that you find the topic engaging and important.

Your audience will read your body language to determine whether you are a confident, credible speaker. They will read the messages conveyed by your posture unconsciously but consistently. They will decide whether you are confident or timid, excited or scared, comfortable or unnerved. Good posture gives the subtle message that you are ready to engage the audience. If you slump to one side, you are saying that you do not care about the topic and wish you were elsewhere.

To start your presentation with the proper posture, your stance should be oriented toward the audience so you can see their faces and they can see yours, allowing good eye contact. At the beginning of the speech, your arms should be at your sides. You may gesture with your arms and hands throughout the speech, but they should start at your sides. Your chin should tilt up, or it will look like you are conceding that you do not know the subject matter. To avoid swaying or rocking during the presentation, you should be leaning slightly forward on the balls of your feet. Knees should be a tad flexed, with your feet located under your shoulders.

With poor posture, you are standing slumped over, gripping the podium throughout the speech with your eyes on your notes, and mumbling so the first row barely hears. It is difficult for an audience to be engaged with any speaker with poor body language. If you suffer from poor posture during speaking, try standing against the wall during a few practice sessions to get used to standing up straight while giving a speech.

Gestures

Beginning a speech with good posture makes it easier to gesture during important parts of the speech. Gestures are the animation of body language; they allow others to interpret context and emphasis while you are speaking. They are made with your hands, shoulders, torso, hips, or any of the body parts that can be moved. During everyday discussions, many people use gestures to communicate, and they come naturally to human communication. You probably aren't aware of the gestures you make because it is an engrained part of your communication style.

It can be hard to adapt your typical communication gestures to the context of speaking in front of a group because speaking to a group of more than four or five people is so different from daily conversation. You are usually anxious about speaking, and that may limit your willingness to gesture. However, gestures are a great help in winning over the audience by keeping them focused on you and by associating the gestures with important parts of the speech.

You may want to make notes on your speech outline of the types of gestures needed to make an important point. They also need to be reflective of the room. Depending on where you are standing relative to the audience, gestures may need to be up near your shoulders for everyone in the room to see. Try to overemphasize gestures so they are not subtle and unseen. Just like projecting your voice to the back of the room, make sure your gestures are large and can be seen by everyone.

Facial Expression

Body language is not as impactful without the actions of your eyes, mouth, and facial muscles that deliver facial expressions. Your face is constantly telling others about your emotional state. Every audience will study your face to determine emotions such as anger, surprise, disdain, or excitement. They want to know how the speaker feels about the subject, and facial expressions are a great way to determine that.

Similar to forgetting to gesture during speaking, an anxious speaker can lose her facial expression and give the audience an unexpressive stone face. A good speaker will unlock her facial expressions with a big smile at the audience to kick off the presentation. If you can incorporate that with an appropriate gesture, even better. The audience feels more at ease, and the speaker has an expressive face right from the start. You cannot smile throughout the whole presentation, but you are unlocking the facial expressions the audience needs to see to supplement the presentation.

Using facial expressions should not be too difficult, since it is an innate human communication method. If you are focused on your message and communicating with your audience, facial expressions should follow.

However, if your face is not communicating during the presentation, you will need to cultivate that skill. There are a couple of different ways you can practice, both involving a mirror. First, practice your presentation in front of a mirror without a sound coming out of your mouth. Your goal is to let your face do the talking. Use your mouth to form the words, but let your face express concern, excitement, or astonishment.

Keep working in front of the mirror to see if you are able to create different moods with your facial expressions. Using the same presentation, instead of trying to show concern during one speech part, use your face to show delight. This is great practice and will prepare you to unlock those facial expressions when you are in front of an audience.

If you want to understand more about the importance of body language, watch the TED talk by social psychologist Amy Cuddy. Not only will you learn about body language, you will also get to see a great speech delivery.

Dealing with Anxiety

One of the toughest aspects of speaking in front of a group is dealing with anxiety or nerves. You are probably familiar with the feeling of dread that blossoms before a speech or the butterflies in the stomach that come from getting up in front of a group of people. I do not know of any sure-fire method for dealing with that anxiety, and I am not aware of any scientific formula that can make it go away. That is because your mind knows you are about to speak in front of a group, prefers that you do something else, and begins working at an unconscious level to signal to your body to react accordingly.

Speaking anxiety takes on different forms that can be manifested when you speak. You may repeat words, unintentionally use filler words (lots of *um*s and *er*s), or stutter. It can also affect your physical delivery because it makes you fidgety, causes you to break eye contact with the audience, or prompts you to pace in front of the audience or indulge in a healthy dose of swaying when you are standing in one spot.

The good news is that everything you are undergoing is a natural response to a stressful situation, which means there are skill sets you can use to help you deal with your speaking anxiety. I am going to offer a few suggestions for dealing with anxiety, including:

- Breathing deeply,
- Visualization, and
- Focusing anxiety.

Deep breathing can be a quick and easy way to deal with nerves right before a speech. If you find that you are getting nervous, take several deep breaths. Focus on breathing in deeply, filling your lungs with air, and expanding your diaphragm as far as possible. Hold the breath for a second and then slowly exhale, letting all of the air out of your lungs. Repeat the deep breathing process as many times as you need until you feel less anxious. You may also want to coordinate deep breathing with warming up for the speech. However, deep breathing to release anxiety is different from the breathing I touched on earlier that provides for clear speaking.

Visualization is another effective technique that can help you deal with nervous energy. The technique is useful in many different tasks, and public speaking is a good venue to use it. The idea behind visualization is to sit down and think about the job in front of you by picturing the task and how it could be accomplished. In the case of giving a speech to the city council, at some point before you are to give the speech, set aside a few minutes in a quiet place. During that time, close your eyes and think about the speech.

Visualize the room where you will speak and the other elements that make up the room. Will there be a podium? How will the room be set up? Where will the audience be located? Am I using slides? Where are the computer, projector, and screen located? How will I start my speech? What are the main points of my speech? How will I end my speech? What will I be wearing? What will my voice sound like? Add any other element you can think of that will affect the speech. As you think about these things, start running the video in your mind about how these things will go.

One of the hardest techniques to master for dealing with anxiety is to channel the tension and energy generated by getting nervous into making a great speech. Instead of focusing all of your thoughts on how scared, nervous, or fearful you are, focus your energy on speech delivery. Let the energy come through your body by changing the tone of your voice, animating your gestures, and showing a passion for the topic.

Do not be afraid to admit that you are feeling anxious. If you acknowledge that you are nervous, it can help calm the jitters. Remember you are there to educate the audience and they are likely to be receptive and considerate toward the speaker. They want to hear what you have to say and are going to be sympathetic listeners.

Do your best to convey confidence to the audience. They are not mind readers and have no idea how anxious you feel. Start with a strong, prepared introduction, and they will react positively to the presentation. As a speaker, you will pick up the good vibe coming from the audience, and it will help calm your nerves.

Keep in mind that all speakers, even presenters who have given hundreds of speeches, get anxious before addressing an audience. That feeling will probably never go away. Practicing will help lessen your anxiety (I will cover that in the section on speaking habits), but even well-polished speakers admit they feel a few butterflies before every speech. It helps if you do not allow your anxious feelings to make you more nervous. Most audiences lend a compassionate ear to speakers, so if you engage with them, they will be receptive and this should help relieve some of the speaking tension.

Turning nervous energy into good, positive energy grows out of your mind-set during a speech. Overcome the dread of giving a speech and embrace the nervous energy generated in your body. Focus that energy on giving an impassioned speech, and you may find that nervous energy will be an asset for public speaking.

* * *

Back to Vivian. Fifteen minutes before noon, Vivian grabs her computer and heads to the conference room. She wants to get her computer set up before her coworkers arrive to listen to her presentation, and she is able to get everything prepped before people start arriving. Her colleagues Rob and Katie have agreed to bring their lunches and listen to her presentation. She also invited Angie, a friend of Tom, to attend. They all arrive within a couple of minutes of each other, and Vivian introduces herself, Rob, and Katie to Angie. She hands them each an assessment form and gives them a little background on why she will be giving a presentation to the city council. When they wrap up that discussion, Vivian asks Rob if he will run the timer on his phone and asks Katie if she will press record on Vivian's audio recorder.

Vivian is a little nervous as she begins, but soon is rolling along through her slides. She is proud of herself when she notices that she is naturally changing the tone and pitch of her voice. She also remembers to pause in a few places and ask the audience if they have any questions. As she concludes, she has Rob and Katie push stop on the phone and audio recorder. Vivian feels that this time was better than yesterday and is curious what she will hear as feedback.

First she asks Rob how long the speech was, and he lets her know it was seven minutes. That is on target for length. She asks them to fill out their assessment forms, and when they are complete, she asks each person to go over the form with her. Rob thought the speech was pretty good, but that there were a couple of places where she went too fast. Katie agreed and added that she had spoken to the council before and found that you really have to speak up in the chambers or your voice will be too soft.

Angie says she liked how the speech was put together, but she tells Vivian she paid close attention to her body language. She could tell that Vivian was anxious because she was pretty wooden throughout the speech. She did not see many gestures or facial expressions that reinforced her message. She noted that Vivian looked at the screen most of the time and rarely made eye contact with the audience. Angie encourages Vivian to pay attention to her body language as she continues to practice her speech.

Vivian is thankful for the feedback. She appreciates their time and has a cookie for each of them as a thank-you for coming to lunch to listen to her speak. She promises to use their feedback to improve the presentation.

Case Study: Mind over Butterflies

Several years ago, I became interested in acting—not as a vocation, but as an avocation: something to do in my free time after I spent my day as an engineer. Getting involved in stage plays probably is not every engineer's idea of a fun hobby, especially an engineer who does not like speaking in front of an audience. However, the idea intrigued me, and I was lucky enough to find an audition at the local community theater with a director who took a chance by casting me in a small role.

During that time, I learned the rehearsal process and enjoyed my role in the production. After a few weeks, the cast was ready to perform the preview night show. This is essentially the last dress rehearsal before opening night and the first before an audience. As we were getting ready for the show to begin, I realized it would be the first time I stepped on stage in front of an audience.

As the time for the curtain to come up came closer, I felt the butterflies in my stomach. The moments ticked away, the anxiety kept building, and I realized that I had not felt that much unease in a long time. If you have a great fear of public speaking, then you know how I felt that night, the nervous energy in your stomach before speaking to an audience.

I cannot say where the inspiration came from, but I began to dispel my concern by thinking about how wonderful it was to feel so much energy. I could feel my nerves in a way I had not felt them in a long time. My senses were heightened, and I understood I was lucky to have this opportunity. I could feel the energy coursing through my body, and I embraced it with a great big smile. I took all the nervous energy and focused on the fact that I was here to do something I had never done before, that I was going to enjoy the experience, and I was going to give a great performance.

One of the hardest things speakers can do is embrace the angst and/or energy they feel when speaking. However, doing so makes public speaking much easier. The next time you are speaking in public and that sense of unease begins to hit you, think about welcoming how you feel and using that energy to focus on giving a good presentation. Flip the nervousness into passion for delivering the best presentation you can.

Experience, preparation, and repetition will go a long way to quell the anxiety you feel before a presentation and help tamp down the tense feelings. However, accepting and harnessing the heightened sensations coursing through your body can be put to use in a positive manner and improve your speaking performance.

Speaking Strategies

While you are working on the skills to become a better speaker, here are a few strategies you can use to enhance your speaking. These approaches help you build credibility and connection with an audience. The strategies of a good public speaker include:

- Appearance,
- Avoiding jargon,
- Engaging the audience, and
- Situational awareness.

These strategies create the conditions that help you succeed. For example, a strategy to match your appearance to audience expectations can make an audience more receptive to what you are telling them.

Appearance

Your look—how you are dressed and groomed—makes a difference to your audience and determines the level of credibility and attention the audience gives you before you have spoken a single word to them. If they see you as someone similar to them or as what they expect an engineer to look like, they are more likely to listen and believe what you are saying.

In Chapter 2, I discussed researching your audience. Exploration of the audience makeup will help you decide what to wear to the presentation. The formality level of the audience will indicate the formality level of your wardrobe. A presentation to a local government body probably requires a different wardrobe than a presentation to high school students.

What you wear to a presentation should be at least as nice as what the audience is wearing, if not slightly better. If you show up at a local government meeting dressed like you just came from the construction site, you may be written off by that audience as a lightweight engineer, and your chance to connect with them may be diminished. However, if you show up to make a presentation to the local contractors group in a full suit and tie, they are likely to perceive you as too stuffy to listen to.

Once you have a good sense of what the audience expects you to wear, find something that is appropriate and comfortable that gives you confidence. Wearing uncomfortable clothing or shoes will take some of your focus off the presentation. Clothes that feel good to wear can help with your confidence and comfort during the presentation.

Along with your clothing, you make an impression with your hair, face, and hands. It is best to keep them clean and tidy. Again, the audience will dictate how important it is to get a haircut before the presentation. If you have a formal presentation, have not had a haircut in two months, and need a shave, get that taken care of before you stand up to speak.

Also in Chapter 2, I recommended having a Plan B for your presentation. If you have an important presentation out of town, consider taking a second outfit. You will appreciate having something to change into if you spill coffee on the clothes you expect to wear when you speak. When you believe the audience is staring at the spot on your shirt, you are not going to be able to relax and focus on your presentation. And a speaker with a big coffee stain on her shirt is less credible than one who is dressed cleanly and appropriately.

Finally, if you have the opportunity, take a quick glance in the mirror before the presentation. Check your clothes to make sure everything is in the right place. Put cell phones (with the ringer off), ID badges, hats, and any other unnecessary items in your computer bag before the presentation so they do not distract the audience while you are speaking.

Avoiding Jargon

Another strategy is using terms and phrases the audience understands. Every profession develops a shorthand language to communicate with others in the profession. There are certain words and phrases within the engineering industry that everyone else who works in the industry understands. However, those outside the profession may not understand engineering jargon.

Jargon is language specialized for a profession that promotes easy communication between two people with a similar understanding of certain concepts. "Walk-through," "valve vault pit," "nonbearing wall," and "expansive soils" are phrases that engineers use every day, but they are not common outside the profession. If you are not careful, it is easy to lapse into engineering jargon during a presentation. Using technical terms as a shortcut during a presentation may seem helpful, but it should be avoided if you want to promote comprehension among nonengineers.

It can be difficult for an engineer to put aside one communication style and shift into another when speaking to an audience of nonengineers. This is true of many professionals, including doctors, lawyers, and accountants. The problem of passing on usable knowledge arises only when they have to communicate outside the profession. How many of you have listened to a doctor speak to another doctor and mostly heard gobbledygook and strange medical terms? Have you ever had trouble communicating with an accountant who expects you to understand complicated accounting procedures?

Probably the other doctors understood and appreciated what the doctor said, but the patient did not. Other accountants understand accounting jargon, but nonaccountants do not. The same can be said of an engineer using technical jargon with an audience on the assumption that the audience is fluent in engineering terminology.

Another problem with using jargon during a speech is that the presentation becomes boring for the nonengineer. Engineers who stick with technical language through the course of the presentation have an amazing knack for taking a subject that is somewhat interesting and turning it into a dry lecture. A few extra words of explanation can help your audience relate to your presentation.

Accuracy is important to what you are saying, but you are much better off speaking in a way that your audience will understand. If you try to impress your audience with your technical vocabulary, you are probably confusing them and preventing them from asking the questions that promote comprehension. Keep in mind that nothing turns off an audience more than a pretentious presenter who is more focused on showing off technical expertise than communicating with them.

A presentation full of engineering jargon can be a real problem when the local government body becomes reluctant to move forward with a project because they do not understand it. Government officials are much more inclined to put a project on hold if they do not understand it because it was explained with complex technical terms. They may also delay the project to get a better understanding of what is trying to be accomplished, and that can be a good thing if they do need to know more. It is better for people to understand complex concepts before they are put into practice. When you are explaining the project to an audience, do your best to make all the terms understandable.

Even the best efforts to replace jargon in a presentation can miss terms and phrases that are not common outside the engineering profession. After the presentation has been designed, find a nonengineer who can look it over for any words or phrases he does not understand. During the presentation practice runs, tell the audience you want them to stop you any time there is something they do not understand. Your goal is to communicate using simple, direct, and helpful terms that make your message easy to remember and share. Once you are aware that engineering jargon is in the presentation, look for ways to explain the phrases with simpler terms.

Engaging the Audience

I've already spoken about the need to engage your audience. But what does that mean? Think of it as the interaction between you and your audience that ensures the audience is interested in the speech from the beginning and stays involved throughout. Audiences notice speakers who connect with them, and they appreciate the outreach enough to stay focused on the topic. Here are a few methods you can use to engage your audience:

- The opening,
- Audience participation and questions,
- Interactivity, and
- Connecting.

The Opening

An audience comes to a presentation to learn something, and they expect to leave more knowledgeable than when they sat down in their chairs. They hope for an experience that is appealing, brings new wisdom, and keeps them informed. They take their seats ready to listen, and from the start a good speaker will give them a compelling reason to stay focused.

A great presentation opens with a strong and appealing foothold. You may not be a born storyteller, but everyone is a born story listener. A presentation that begins with a story will hook your audience with an emotional connection. In Chapter 3, I covered how to begin your presentation with an engaging story or interesting fact. Remember, it does not have to be long or funny, just relevant to the topic. The more interesting you can make the opening of the presentation, the more likely the audience is to stay involved.

Once you have the audience paying attention, signal to them what the presentation is going to cover. You can do this overtly or a little covertly, but let them know the speech's goal and your expectations for what they will learn. When you foreshadow the major points in your speech, your listeners can better follow along and they can gauge how far along you are in the speech. Grab their attention at the beginning so that they stay with you to the end. Lyndon Johnson said, "If they're with you at the takeoff, they'll be with you in the landing."

Audience Participation and Questions

Your audience is more likely to stay engaged when there is some back and forth between you and your listeners. As part of your preparation, consider how you are going to let your audience participate during the presentation.

- Will they be able to ask questions throughout?
- Should they save their questions until the presentation concludes?
- Will you ask them for input at certain points along the way?

Few presentations do not generate audience questions. The more opportunities you give an audience to ask questions and participate, the more engaged they will be with your presentation. It also helps to ensure that the audience understands an important point before you move on to the next one. If it is not appropriate for the audience to ask questions during the presentation, be sure to use other techniques to keep their attention.

There are some dangers in allowing audience questions during your presentation. If the questions get too far away from the main topic, they can derail your presentation. If you see something like this happening, tell your audience that you'd like to finish your presentation and return to the questions at the end. If you have limited time and need to get through the material in a timely fashion, then preemptively ask the audience to wait until the end for questions. If you have a topic you expect will generate a lot of audience questions, but you expect to answer those questions during the presentation, again, ask the audience to wait until the end to ask questions.

If you are taking questions from the audience throughout the presentation, consider inserting a black slide in the slide deck to indicate when you want them to ask questions. A black or blank slide will force the audience to shift their attention from the screen to the speaker, allowing the speaker to address their questions.

Interactivity

Another way to enhance audience engagement is to make them an active part of your presentation. Here are a couple of ways to do that.

For instance, if you ask the audience questions about your subject, give the answering person a small gift (a candy bar or company pen). Once others in the audience know they may be rewarded for paying attention, they are more likely to do so. Try giving the audience a short quiz and challenge them to write down answers that reinforce the presentation topics. Getting the audience to participate in a physical activity—even a simple show of hands—is a great way to generate energy and keep audience members alert.

Connecting

A good speaker makes a personal connection with an audience. You want to establish a synergy between you and the people listening to you.

If your audience believes that you are showcasing your authentic personality, they will be open to a connection. Good speakers use and hone their public speaking skills to communicate with their audiences, while allowing their individuality to shine. Don't force yourself to be someone you are not because you think that is what the audience wants; you will probably come across as too polished and artificial.

Allow your personality to be seen and felt by your audience. When an audience senses that you are genuine, they have a much easier time trusting you. They want to see your passion for the topic. They may not understand every concept you cover, but when they trust that you believe what you are saying, they will connect and engage.

Situational Awareness

Understanding the situation in which you are speaking is another strategy used by good public speakers. Each presentation has its own context, and you will want to anticipate the conditions and adjust your delivery accordingly. Having a sense for the mood of the audience is always helpful. The physical environment or the particular issues of the day can help you predict how an idea or project will be received.

Let's say you are slated to give a presentation to a local government body. You get a copy of the agenda and find out your project is one of the last items. If several contentious items precede your slot, you are likely to face a council that is tired and in a poor mood. So, good situational awareness might suggest that a short and sweet presentation with a willingness to answer questions would be well received. The last thing the council needs at the end of a protracted meeting is a long-winded engineer.

A speaker talking to a local service club after major events in the community should be ready to address questions related to the topic. If you are slated to talk about storm sewer improvements the week after flooding destroyed several homes, you are going to be asked about the flooding. You should adjust the presentation to address what is going to be on the minds of the audience.

You risk looking callous and unconcerned if you are giving a presentation on a new road that will provide a speedier way to get across town during a month when a young doctor was killed by a drunk driver, a child was killed in a school bus collision, and a major car wreck put several people in the hospital. A presentation after this series of events will be seen in the context of all that happened the month before. If you aren't aware of related events, you can't address them directly and you risk losing your audience's trust.

Time of day can also make a difference. Speaking at a technical conference in the morning is not the same as giving a presentation in the afternoon. The audience is usually alert for morning presentations and groggy for presentations just after lunch. If you are speaking after lunch, you may need more audience-engagement strategies, for instance.

Situational awareness during a presentation allows you to prepare for questions that will arise, be knowledgeable about the situation, and understand how the audience is going to react. It allows you to shorten or expand your speech as needed. If you are presenting in the afternoon, ask the audience to stand and stretch if they seem sleepy after lunch. A presenter who has been following the news, can see how it relates to the speech topic, and adjusts the speech accordingly will have a much better chance to communicate effectively.

* * *

Vivian is able to schedule a time for Tom to watch her practice the presentation. They are in the conference room, and she is feeling much more comfortable now that she has practiced the speech a few times. She gives Tom her phone to record the audio and tells him to start the timer when she begins. Vivian is working on incorporating some movement into her speech, and now that she is more practiced, she is more at ease giving the presentation. Her voice changes and gestures are more pronounced, and she moves smoothly from one slide to the next.

She wraps up the speech and asks Tom how long it took. Six minutes and 45 seconds, right where she wants to be. Vivian asks him if they can go through the assessment sheet, and he agrees. Tom thought she did well with her voice and movements, but he did add that she could probably spice up the beginning a little bit. He gives her some ideas for a few quick facts she could add to the beginning that should get the city council's attention. Tom notes that she did a good job of finding spots where the audience can ask questions, but he reminds her not to allow the questions to meander too far off topic. He also tells her that he has heard the road project on the west edge of town is not going well and she should at least read the stories in the newspaper about the problems between the contractor and the city.

As they start to pack up, Tom asks Vivian how many more times she is going to practice her speech. She lets him know that she is planning on spending an extra half hour after work in the conference room running through the speech at least two more times. Tom knows this is Vivian's first time presenting to the council and wants to provide a safety net. He lets her know that he is planning to attend the meeting, sitting in the back of the room, and if something comes up that she does not anticipate, he will be there to help her.

Case Study: Technical Proficiency versus Communication Skill

The greatest virtue of good engineers is their technical proficiency, right? Who would want to hire an engineer who is anything but the most technically competent they can find? It turns out that most people who hire engineers are not interested only in their engineering skills. Clients find other qualities much more important.

PSMJ Resources (www.psmj.com) polled a number of public works departments about the engineers who work for them. They asked them to list the most important skills engineering project managers must have to do a good job. Here is the list they came up with:

1. Follows through on commitments,
2. Good listener,
3. Proactive,
4. Nails every aspect of the job,
5. Leads by example,
6. Good communicator,
7. Backs decisions of team members,
8. Organized,
9. Handles multiple priorities well,
10. Technically proficient,
11. Holds people accountable, and
12. Delegates well.

The survey shoots a hole in the bucket of any engineer who believes that technical skills are all that is important to being a good engineer. An engineer who wants to be an above-average project manager should be practicing communication skills as much as, if not more than, engineering skills. Soft skills—communication, project management, working with people—should exceed engineering skills if you are going to stand out as a project manager. The emphasis on developing soft skills is a highlight of ASCE's *Civil Engineering Body of Knowledge* (2008), which outlines what is expected from a competent engineer.

A review of the list shows how crucial communications skills are to excelling as an engineer. People working with engineers believe skills 2 and 6—listening and communicating—are higher on the priority list when hiring a project manager than technical proficiency, listed 10th.

The communication skills rated in the survey are helpful to becoming an accomplished public speaker. A good listener will be able to determine what an audience needs to learn from a presentation and incorporate that information into her speech. Effective speakers will take that material and communicate it well to the audience, in a manner that promotes understanding. Engineers who communicate well are a sought-after commodity, and, for a project manager, the ability to communicate with the client is more valuable than technical proficiency.

A good engineer needs to be technically proficient, but the people working with them appreciate (and hire) engineers who communicate well. If you are wondering if it is worth your time to practice your public speaking skills, the PSMJ survey should reassure you that the investment in improving your talents is worth it.

Speaking Habits

Nothing helps improve your public speaking skills like practice. It is the *habit* of public speaking that matters and reinforces your skills and strategies. Repetition and deep familiarity help polish the presentation, bring out the best speaker in you, and moderate speaking anxiety. The more engineering projects you work on, the better engineer you become; the more experience you have giving speeches, the better public speaker you become.

As you invest in becoming a better public speaker, you will notice improvement over time. After you deliver (and survive) a less-than-stellar speech, you realize the experience is not so bad and you will actually have more confidence in yourself as a public speaker. By the time you have spoken to audiences several dozen times, your confidence will grow and you will become a good, or even great, public speaker.

To get there, you must foster the habit of practicing your presentations through

- Focused practice,
- Practicing out loud,

- No substitutes for practice, and
- Finding where to practice.

Focused Practice

In all aspects of life, practicing a skill makes you better at it. The more you practice engineering, the better engineer you become. To get better at speaking in public, you are going to have to practice public speaking. Nothing does more to put you on the road to becoming a better public speaker than focused practice. Nothing lessens speaking anxiety like preparation.

It is easy to say you will commit time to practice speaking. However, it is also easy for practicing to go on the back burner as other priorities arise. All of us are busy people, and you will be tempted to focus your time and attention on skills you consider more essential to engineering than public speaking. What you may miss by not practicing speaking skills is the chance to share engineering knowledge with an audience.

So if you are committed to improving your public speaking skills, you must set aside practice time. How much time will vary, depending on the skills you need to improve and the importance of the presentation. If you have a great deal of anxiety about public speaking and an important presentation looming, you may need 8 to 12 hours of practice time for a 30-minute speech. For a less important 30-minute speech, you may need to set aside only an hour and a half to two hours. Time spent on the planning and design is not included in practice time. Practicing your speech means time spent practicing out loud. At the very least, a speech needs to be practiced out loud three times.

Practicing Out Loud

Focused practiced is a good start, but it's not enough. You must also *practice out loud*. This is the key to pulling together all the elements I've talked about in this book. It is the trial run of your presentation and the best way to discern how the presentation will turn out. Being prepared to speak to an audience means you have actually practiced delivering your speech. It requires standing up and enacting the presentation by speaking the words out loud. It can be done in an empty room, in the room where you will deliver the speech, or to a friendly audience.

Before I get into the details of practicing out loud, I want to be very clear about what practicing out loud does *not* mean:

- It does not mean going over the project technical details 1,000 times before the presentation and being overly prepared to answer the smallest detail about the project's engineering aspects.
- It does not mean sitting at your desk reviewing the slide deck.
- It does not mean silently reviewing the presentation, saying the words in your head.

When you are practicing out loud, if someone were to walk in the room, she would be able to hear what you are saying.

It is easy to fall into the trap of preparing for the presentation by going over the technical details until you can answer every conceivable question about any design aspect. Most likely that is the reason you became an engineer. You would much rather spend time working on the technical details of a project than speak to a group of people about the project. As a competent engineer, you should know the project's technical aspects and should be able to answer any questions that may arise from the audience.

Practicing out loud probably does not sound like a lot of fun, and you will be tempted to practice by looking over the slides and repeating the words in your head. You will sound great and no one will be able to criticize you. However, you will not be able to hear the volume, tone, and pitch of your voice, the stutters and filler phrases as you search for the right word, or where you need to take a breath to get your voice back. Without practicing out loud, you cannot accurately determine the speech's length.

The great thing about practicing out loud is that you can make all your mistakes before you have to speak in front of people. Instead of finding out about your poor vocal variations from the sour looks on the faces of the audience, you get to hear it in the audio recording so that you have a chance to fix it. You learn what presentation sections you struggle with, which are shorter than you expect, and when you are going to want to move around during the speech. You get the chance to fix these problems by practicing instead of allowing the audience to see you struggle through them.

You should practice delivering the speech out loud several times, no matter how much you dislike the exercise. Practicing out loud means getting up on your feet and speaking in your speaking voice, as if people are there to hear you. As discussed earlier in this chapter, the physical speaking skills (voice, tone, posture, and so on) are important to speech delivery, and you have to use them to know how they are working.

Find a desk or chair that can substitute for a podium and move around as though you are presenting in front of an audience. Set a timer, such as the one on your phone, and begin delivering your speech with the same volume of voice you expect to use when you are in front of the audience. Try to have an audio recorder, like the one on your phone, or a video recorder going so that later you can review how the practice session went. The only way you are going to be able to assess and improve speaking skills like tone, volume, speed, pitch, and breathing is practicing your speech out loud.

No Substitutes

A word of caution: do not think you can read this section of the book and easily use the skills and strategies as a shortcut to becoming a proficient public speaker. If you think the next time you speak in front of a group you will do a good job with eye contact, change your voice control, dress to look like an engineer, dumb down the speech so no engineering language is included and everything will be just fine, you could be disappointed in how things turn out. Except for a few speaking prodigies, everyone else has to work to master the art of public speaking. While I encourage you to try the suggested techniques, you have to do so in conjunction with plenty of practice. No amount of mastery of technique can make up for the experience of practicing the presentation.

Please do not fall under the illusion that practice will make you the perfect public speaker. Instead, focus on achieving the goal of becoming a practiced, competent, effective communicator with the audience. Practice does not make perfect ... it makes us better prepared. By no means am I saying that you must be a polished public speaker before you speak to an audience. However, it will help immensely if you continue to work on your skills and strategies before and during your public speaking engagements. In fact, one of the best ways to put your skills to the test is by speaking at public meetings. It is the best practice you can get.

Improving your speaking skills may take years of practice. If you begin as a poor public speaker, it is important that you get better over time. Do not expect results overnight or over the course of preparing for one presentation. You will need to work on your skills through several presentations. If you are a good public speaker, there is always room for improvement. You do not have to be a bad public speaker to improve.

Finding Where to Practice

After imploring you to practice public speaking, I wanted to provide a few opportunities that are available to practice public speaking. If you are not a seasoned public speaker, you could have some trouble finding places to speak to an audience. Be sure to ask the people you work with about public speaking opportunities available to employees. Many companies encourage their employees to join Toastmasters and will provide the resources needed to participate. Here are some other ideas.

Lunch and Present

If you are looking for a venue to practice speaking skills and you work in an office, ask the boss if you can schedule a regular time and place to practice presentations. In a large enough office, people are speaking on a regular basis, so set up an opportunity for speakers to give their presentations to coworkers every second and fourth Friday of the month. With enough people in the office giving presentations, you can find different speakers willing to speak on a Friday. The meeting could be over lunch in the conference room. You can boost attendance by providing lunch that day. Find a method to ensure that the people who show up give feedback on the presentation.

If you work in a small office and a regular meeting is not feasible, you can set up the same type of event and ask people to bring a brown bag lunch. Have coworkers invite their spouses for lunch to get people other than engineers to listen to the presentations. Tell them you have to give a presentation and need feedback from nonengineers. Bring a thank-you for the people who do come to listen (such as cookies or pumpkin bread).

Several variations on this type of idea provide a venue for people at work to practice their public speaking skills. Find the one that works for your office and schedule some practice sessions.

Volunteer to Give a Safety Presentation

Many organizations have a workplace safety program that includes routine safety presentations. Ask the safety coordinator if you could make a presentation on a safety topic. If the coordinator agrees, go through the process of researching a subject and developing a presentation. Try to videotape the presentation so that you can review it, and collect evaluation forms from the audience.

Service Clubs or Speaking Classes

Each morning and noon hour across the country, thousands of service groups (such as Rotary, Exchange Club, Kiwanis, and chambers of commerce) are meeting, and they are often in need of speakers. Contact the local groups and volunteer to give a speech to their organizations. The topic should be of some interest to them, so if you are working on a big public works project in the community, they will likely be interested in having you come give a speech on the subject. Ask the project manager to ensure it is okay to speak about the project, and if he agrees, you've got a great way to get in front of a pretty sympathetic audience and practice public speaking skills.

There are classes available to learn public speaking techniques. Seek out an organization that specializes in training public speakers or find out if the local community college has a class on public speaking. There are clubs that work on public speaking on a regular basis. The mission of Toastmasters (www.toastmasters.org) is to help people improve their public speaking, so find a club that meets near you and go to a meeting.

Speak When No One Else Will

Meetings of local government groups—town councils, zoning boards, planning or advisory boards—provide many opportunities for the average citizen to speak up and be heard. For instance, here are a couple of agenda items:

> Resolution relating to revenue bonds for the 2014 and 2015 wastewater replacement projects; authorizing the issuance and fixing the terms and conditions of the bonds.

and

> Resolution relating to pooled series 2015 special improvement district bonds; fixing the form and details and authorizing the execution and delivery.

Those are actual agenda items from a meeting that took place while I was serving on the city council. As you might imagine, no one came to testify on either of them. That does not mean there shouldn't have been people to advocate for those projects. Both items passed easily, but they were also opportunities for citizens to speak up.

If you are looking for a place to practice, items such as these may be just the ticket. When you have no professional or other ties to a project, the public hearing is a great place to gain experience. Take advantage of opportunities to speak on infrastructure projects. Prepare for the speech as you normally would, and be willing to answer questions.

If you are unsure what to say during such a hearing, speak about the importance of infrastructure in a community. It will be good practice, it will help highlight the importance of infrastructure in the community, and it will be good for the council to know that there are people who care about infrastructure.

Volunteer to Speak about Qualification-Based Selection

Consider reaching out to a local government to ask about giving a presentation on the Qualification-Based Selection (QBS) process for selecting and hiring engineers. Many local governments use the QBS process, but many local elected officials do not have a great grasp of the QBS process. In my estimation, about 10% of council members do not understand QBS; about 30% think it is a great way to choose an engineer, and about 60% think it is a poor way to select an engineer.

Most local governments are used to paying for things using the low bid system, so individuals may have difficulty adjusting to the QBS process. A presentation from a knowledgeable professional (and colleague) can help them get a better understanding of the QBS process. The presentation can be timely if you notice that the council is having trouble with QBS or that they are going to be selecting engineers for upcoming projects. You could speak by setting up an informational session for the governing body at one of their informal meetings. Find out how you can get on the agenda by contacting the staff person who sets the agenda. Or, at a regularly scheduled meeting, tell the council you would like to discuss QBS and ask them when you can make a presentation. A third option is scheduling individual appointments with each local elected official to discuss QBS.

Career Day

Most high schools set aside a day during the year to have professionals speak about their jobs. Call the local high school and ask if they have a career day and whether you can speak about being an engineer. Not only is this great practice, but you'll be exposing young minds to the engineering profession. Young folks are a very honest audience: they will either be engaged or they will not. Either way, you'll learn a lot!

Case Study: A Missed Opportunity

It was Engineer's Week, and I was able to attend a nice banquet at a big hotel celebrating the week. The evening commenced with drinks and socializing among local engineers and their guests, followed by a decent dinner. The evening's program began about halfway through the meal, with the emcees telling some bad jokes and handing out a few door prizes. As dinner wrapped up, scholarships were given to local high school students headed for engineering schools. The engineering outreach events from the past year were highlighted, and a new batch of Professional Engineers were given plaques after reciting the Engineers' Creed. A heartfelt moment was shared when a new member of the Engineering Hall of Fame was introduced and reflected on his career.

The evening went along swimmingly until the Project of the Year Awards. All the goodwill and energy built up over the course of the evening was quickly drained by back-to-back dreadful speeches from recipients of the Project of the Year Award. It was like an iron curtain dropped to halt any of the evening's enjoyment. The first engineer-speaker received the small project award. He was unprepared to speak about the project, rambled off the subject, and could not make himself heard. It was dreadful.

The next speaker was receiving the Project of the Year Award for a large project. Again, the presenter was unprepared and made an important project in a significant area seem extremely bland and uninteresting. The audience was figuring out how to look graceful when charging to the exits instead of listening to the speech. Who could blame them? These two speakers ripped the energy from the room because they could not be bothered to prepare adequately for an important public moment in their careers.

When engineers fail at public speaking, they are missing valuable opportunities to provide people with an understanding of the important work that engineers do. You could argue that the banquet was attended by a bunch of engineers, so an effective public presentation was not needed to highlight engineering. But the banquet was also attended by spouses who probably did not understand many aspects of engineering but might have wanted to. The speakers missed a chance to highlight the cool engineering projects for a group of students who were there as winners of a prestigious scholarship to study engineering. Finally, the speakers missed an opportunity to inspire the students' parents, who would have been proud to see what their kids could become if they pursued engineering.

These speakers had been notified in advance that they were receiving awards, so they could have made the effort to prepare a presentation. They chose not to prepare and gave terrible speeches. What a missed opportunity!

---

* * *

The moment has come for Vivian's big presentation to the city council. When the meeting began, the room was a bit chilly and the air was fresh. After three and a half hours, though, with a packed crowd of people waiting for a turn to speak, the atmosphere is stuffy, warm, and stale. The council meeting has already dealt with two contentious hearings, the first a proposed ordinance to limit dogs in city parks, the second about a commercial development that will have a big impact on a well-kept, long-established neighborhood. Both hearings had several people testify on either side of the issue and close votes—banning dogs that weigh more than 85 pounds from city parks and allowing the developer to move forward with the commercial development.

After a 10-minute break, it is Vivian's turn to make her presentation to update the council on the road project. She was 15 minutes early to the meeting, so her presentation is loaded on the computer and projected on the screen as she gets up to speak. Since this is her first time in front of the city council, they do not know what to expect from her and are paying attention as she begins to speak. She notices Tom sitting in the audience and it helps with her anxiety. She opens with a few facts about how the underperforming road is affecting the town, and it is clear that she has hooked the audience's attention. She provides them a preview of what she will cover and launches into the meat of the speech.

Vivian covers budgets and time lines, then puts up the graphic on project beginning and ending limits. She lays out the next steps in the process and tells the council that she will need their input on several issues, including the type of road cross section they prefer, whether or not a trail crossing should be added to the road, and depending on the results of those decisions, how much money they are willing to spend on the project. They ask her a few questions, and she handles them easily. She wraps up her presentation and tells the council she is willing to keep answering questions if they have any. The mayor looks at the other council members, and one of the council members raises his hand and asks how the county commissioners responded to the presentation. Vivian replies that she has not briefed the county officials on the project. The council member seems satisfied with that answer and without any more questions arising, Vivian thanks the council for their time and sits down.

The public works director stands up after Vivian and says he needs council input on the three items that Vivian discussed. He requests that the mayor have council members weigh in on each issue. They pick the cross section, decide for a trail crossing, and agree to allocate more money for the project. With those items taken care of, they move on to the next agenda item.

Vivian is able to leave the council chambers knowing the next steps in the project and can get started on them. She meets Tom in the hallway outside the chambers. He smiles and says, "You did a great job, Vivian. However, I did want to talk with you about the question you got about the county commissioners."

References

ASCE. (2008). *Civil engineering body of knowledge for the 21st century.* <http://www.asce.org/uploadedFiles/Education_and_Careers/Body_of_Knowledge/Content_Pieces/body-of-knowledge.pdf> (Feb. 21, 2017).

Chapter 5

Local Government

Every two years the American politics industry fills the airwaves with the most virulent, scurrilous, wall-to-wall character assassination of nearly every political practitioner in the country—and then declares itself puzzled that America has lost trust in its politicians.

—*Charles Krauthammer*

In this chapter, you'll learn about another piece to this communication puzzle: your local government. First, you'll get an overview of the structure of local government, how it works, and who the key players are. Then—because for any project to get designed and built, someone has to pay for it—you'll be introduced to the funding mechanisms used by local governments to pay for their infrastructure. I will discuss how to become familiar with the people who play important roles in local government, as well as how they conduct business. Finally, you'll be inspired (I hope) to get involved with your local government and even become an advocate for infrastructure.

This chapter has three case studies that delve into the role of city council members and engineers. The first offers some insight into what local elected officials work on during their tenure. The second covers what can happen at a local government meeting when an engineer relies on technical knowledge rather than presentation skills. Finally, I will show how the local government process can go awry when people are not paying attention to it. An appendix to this chapter lists several resources for more information on local governments.

Getting to Know Your Local Government

Having an interest in how Vivian's speech turns out and wanting to make sure she succeeds, Tom attends her presentation at the city council meeting. He is pleased to see that Vivian does a great job on her speech and that he does not need to be called on during the meeting. As she presents, the council asks several good questions. He knows this means she communicated well and they paid attention to what she had to say. The last questioner asks how the county commissioners responded to the presentation. Vivian replies that she has not briefed the county officials on the project.

Tom could tell from Vivian's face that she was perplexed by the question, so he follows up with her the next day. She says that she was confused because she thought only the city would be interested in the project. Tom points out that things are rarely that simple. The county will want an update on the project because a portion of the funding for the road comes from dollars that are passed through the county. Moreover, the county has plans to upgrade the road section in its jurisdiction the year after the project is completed in the city. Vivian still looks puzzled.

Tom explains that city road ownership ends at Dryer Avenue, and from there the road belongs to the county. Vivian did not know that the county owned and maintained any roads so close to the city. Tom decides that Vivian is ready for another lesson in real-world engineering: she needs to understand how local governments interact if she is to understand how road funding works. He grabs a sheet of scratch paper and sketches out the local government hierarchy. Tom explains how the city council relates to the county commission, and Tom adds a few personal war stories about working through the process at the county and with a former city council member whom he really liked. By the end of the discussion, Vivian is better informed about local government and is interested in learning more.

Vivian's understanding of local government is similar to that of many engineers because local governments are the least understood level of government. However, it is crucial for engineers to understand who has authority and who provides funding for infrastructure projects such as roads, bridges, airports, schools, water and wastewater treatment plants, and landfills. An engineer who understands how local government functions will not only give successful presentations, but will also be effective working with those governments.

Government Hierarchy

Governments in the United States are set up in a tiered system. At the top is the federal government, the next step down is state government, and the lowest rung is local government. The U.S. Constitution does not specifically mention local government entities; rather, the system of local government is created and regulated by the states. Local government jurisdiction covers smaller geographic areas and is the level of government closest to the people. Because each state designs its own system of local governance, there are vast differences among local governments from state to state.

This is a good time for some statistics. The Census Bureau is responsible for collecting statistics on governments in the United States, and conducts the Census of Governments every five years. In the 2012 Census of Governments, the Census Bureau identified a total 89,004 local governments. Separated into the five Census Bureau categories, there are

- 19,522 municipal governments,
- 16,364 town or township governments,
- 3,031 county governments,
- 12,884 school districts, and
- 37,203 special district governments.

That's a lot of local government spread across the United States, and at some time or another many will engage the services of an engineer to help them design infrastructure. Engineers work most frequently for public works projects sponsored by cities and counties. Let's take a closer look at some of the flavors of city and county governments.

City Government

Historically, the most common organizations of city government are mayor and council, council with a manager, and commission or representative town meeting.

Mayor and Council

The mayor–council form of government is characterized by a separation of powers between the directly elected mayor and precinct-elected city council. Generally, the mayor has executive powers, while council has the legislative powers. Often the mayor job is a full-time position that includes compensation.

This form tends to exist in older, larger cities, or in small cities, with a population fewer than 25,000 people. Cities with variations on the mayor–council form of government include Los Angeles, Chicago, Topeka, and Minneapolis.

Depending on the city charter, the mayor could have weak or strong powers. The strong mayor is most often seen in big cities, such as New York, Chicago, and Los Angeles. In these cities, the mayor runs for office, is elected, has a full-time position, and controls the city management. Strong mayors are typified by their central executive decision-making authority and the ability to hire and fire the staff working closest to them (normally department heads like the police chief and fire chief). They have a veto power over the city council, and the city council has little to no authority over day-to-day actions of the city. Small towns may also use the strong mayor form of government because they do not have the resources to pay a manager for a handful of city employees, so they put the mayor in charge. In those smaller cities, being the mayor is not a full-time job and that individual usually holds other jobs in town or is retired.

Local government with a weak mayor is commonly referred to as the council–manager local government. A weak mayor serves on a city council that holds all of the power and authority. Weak mayors are unlikely to supervise staff, are not able to hire and fire staff, and do not have veto power over actions of the city council. They serve mostly as figureheads, presiding over meetings and breaking tie votes, but they do not have more authority than any other member of the council.

Council with a Manager

A local government formed around the council–manager structure has a city council to oversee the general administration, make policy, and set the budget. It is common to see the council–manager form of government in midsize to smaller cities such as Phoenix, San Diego, Salt Lake City, and Rockville, Maryland. If there is a mayor, the incumbent is chosen from among the council members on a rotating basis. The mayor's role is usually a ceremonial position with no powers over and above other council members.

In this form of local government, the council hires a city manager to carry out day-to-day administrative operations. The city manager is a skilled administrator hired to manage a decent-sized organization. City managers are expected to maintain political neutrality. The mayor and city council oversee the manager, with the ability to hire and fire the manager.

Commission or Representative Town Meeting

Less often-used arrangements for local government are the commission and the representative town meeting. In the commission form of government, voters elect individual commissioners to a small governing board. Each commissioner is responsible for one specific aspect, such as fire, police, public works, health, or finance. One of the commissioners is designated as chairman or mayor, who presides over the meetings. The commission has both a legislative and executive function. The commission form of local government is the oldest form of local government, but it exists in few cities today.

The representative town meeting is primarily practiced in the northeast part of the country. Town meetings are acclaimed as the purest form of democracy because all voters have a direct say in how the town is run. However, it is usually effective only in small towns and is a very representative form of government. Each town meeting must be announced with a date, time, meeting location, and the items to be discussed. Individuals vote on town legislative matters at the town meeting. Selectmen are elected to implement certain policies decided at the town meeting and to manage the affairs of the town in accordance with state statutes.

County Government

County governments are generally run in one of three ways: by a commission, by a commission and administrator, or by council with a county executive. While not everyone in the United States lives in a city, everyone does reside in a county.

Commission

Under a county commission, the power to enact ordinances, adopt budgets, administer policies, and appoint county employees is delegated jointly to an elected commission. Although governing body members are most usually called commissioners or supervisors, these are not universal titles. Some governing body members in Louisiana, for example, are called parish police jurors. The county governing body in most New Jersey counties is the board of chosen freeholders.

Commission and Administrator

Under this form, the elected county board of commissioners appoints a paid, full-time administrator who serves at its pleasure. That individual is in charge of a wide range of powers, including the authority to hire and fire department heads and formulate a budget. Administrators are hired to provide professional management across all parts of county government.

Council and Executive

The separation of powers principle undergirds this governance system. An elected county executive is the paid, full-time chief administrative officer of the jurisdiction. Typically, the county executive has the authority to veto ordinances enacted by the elected county board and to hire or fire department heads.

Although a majority of counties still operate under the commission form, more than 40% are being managed by an appointed county administrator or elected county executive. State policymakers have contributed to this trend, as Arkansas, Kentucky, and Tennessee now mandate that counties in those states be headed by an elected executive.

Special Purpose District Governments

One other type of local government that may require the services of an engineer is a special purpose district. There are hundreds of different kinds of special districts, which are independent government units that are not cities or counties. They exist autonomously, often with their own appointed or elected board. A special district is often modeled on city or county governments and can have its own staff and administrative personnel. A special district is put together for one purpose and serves the people who live in the district for that one purpose. Examples of some special districts include airport authorities, port authorities, cemetery districts, irrigation districts, and water and sewer districts.

Case Study: A Day in the Life of a Local Elected Official

Having served as a local elected official, I can offer a little perspective on the tasks these officials undertake, the pressures of the office, and the variety of issues they confront. Let's look at some of the typical tasks encountered by local elected officials.

The first rule of a typical day in the life of an elected council member is that there is no such thing as a typical day. Council members are tasked with vast and various duties. They need to be prepared for meetings on land use, budgets, and infrastructure. They should understand the role and purpose of all the city departments, seek input from citizens, provide oversight and guidance to city staff, and interact with the organizations that are affected by city decisions. On the plus side, council members get to hear about something new every week. On the minus side, they usually have little time to focus on any particular issue before the next one comes along.

Like the vast majority of city council members in the United States, my job as a member was not a full-time position. My full-time job is being a water and wastewater engineer. Therefore, I did my council work during the times I was not working at my full-time job. I was a part-time council member because my city has a council–manager form of local government. As one of eleven elected representatives, I helped set city policy objectives. Once those objectives were agreed upon, the policy was administered by the city manager.

The council's work included setting policy (such as making difficult decisions about a policy not everyone agrees with), making land-use decisions, aiding constituents with problems related to city government, and formulating a city budget. In my city, the council makes decisions regarding the airport, transit, parks and recreation, police, fire, public works, administrative functions, and planning. At any particular meeting, the council may award contracts at the airport, set the water and wastewater fees, decide the animal shelter's fate, and determine if a large parcel of land will be annexed.

On top of the formal duties of a local elected official are the many informal duties of a council member, which often take the shape of meetings with various groups or individuals. To stay informed and keep a finger on the pulse of the city or county, I found that these are great meetings to attend. The meetings run the gamut from soup to nuts on significant community issues. As an elected official, I knew it was important to interact within the community and listen to the issues that are raised during the meetings.

I hope you can see that the daily routine of a local elected official involves a vast array of activities and issues. If local officials are going to be the audience at your next presentation, take the time to understand their viewpoint: they are being asked to make the best decisions they can for their communities on a variety of contentious issues that may involve highly technical subjects in which they are not experts. It's up to you to make them trust you to be their technical expert.

Funding Mechanisms

Everyone likes a new infrastructure project, but no one likes paying for it. So a major hurdle for local governments is how to pay for design and construction. Traditionally, local governments had two options: pay cash or borrow money. Over the past few decades, a couple of new options have become more widely accepted, and many municipalities are using alternative funding sources, such as public–private partnerships and design–build–operate, to move projects from conception to reality. Keep in mind that local governments may use several of the financing options available to them to complete their infrastructure projects.

A local government can save up the amount it needs to pay for a project by collecting taxes or by being issued grant funding. When the money is in place, it can move ahead with construction. This method of payment is called *direct payment* because the local government raises revenue through fees, taxes, or grants to pay for the job up front. Once construction is complete and the contractor paid, the local government owns the infrastructure free and clear. It will pay for operation and maintenance, but no debt is associated with the project—that's the biggest benefit to this option.

A local government unit may also create a new special district specifically for the purpose of building infrastructure. A sidewalk district could be used to build sidewalks within the district boundaries. A portion of the taxes collected in that district will be used to directly pay for sidewalk design and construction. Another direct payment option is to require a private party to pay for the improvements. This is often associated with land development. When a company proposes to build a large subdivision that will have an impact on the roads near that subdivision, the local government may require the developer to pay the cost of upgrading the roads, water and wastewater system, and so on. This way, the local government does not have to pay for the infrastructure improvements, but will take ownership of the improvements after they are built.

Alternatively, a local government can finance the project with some method of debt. This payment method is called *debt financing*. The local government borrows money to pay for the project design and construction. Once the project is complete, the local government is obligated to make a monthly or yearly loan payment. The loan is paid with revenue generated from taxes and fees, but the money does not have to be available up front. The biggest benefit of this option is that infrastructure can be built and used while it is being paid for; it can even be partially paid for by new revenues generated as a result of the project.

A plethora of options are available for debt financing by local governments. While not an exhaustive list, some of the funding sources may be:

- General obligation bonds,
- Revenue bonds,
- Tax increment financing,
- State revolving funds, and
- Infrastructure banks.

A popular debt financing option is to use general obligation (GO) bonds. These bonds are backed by an assurance from the local government that it will levy the taxes necessary to make the payments on the bonds. A GO bond is a loan taken out by a local government against the value of taxable property in its jurisdiction. A bond bank is set up at the state level to allow smaller local government units to get better terms than borrowing on their own. Small cities may not be able to obtain bonds on their own, but if they pool together with other small cities across the state, investors are more willing to lend them money. The bond bank is able to offer lower-cost loans because the statewide bank provides more stability.

A local government may use a revenue bond to pay for infrastructure if it knows that the project will generate money in the future. A toll road paid for by a revenue bond will pay back the investors with income generated by the toll road. Revenue bonds are different from GO bonds because only the revenues from the project are used for payback, and the local government is not required to use other tax money to pay back the bonds.

Local governments that use tax increment financing do so by first creating a district with defined geographic boundaries. Once the district is created, a baseline of property tax revenue is established. As the property tax revenue increases over the years, the amount of the increase over the baseline (the increment) is used by the district to invest in projects within the district. These are often infrastructure projects that benefit the residents and businesses in the district. This allows the local government to invest property tax revenue from the district back into the district without having to share that revenue with other governments.

Another popular option is the use of revolving loan funds or infrastructure banks. A typical revolving loan fund is a state or federal program that has a pot of money to lend local governments for projects. The loan funds rely on principal repayments, bonds, interest, and fees to recapitalize and replenish the fund as a perpetual source of debt financing. An infrastructure bank selects projects for funding based on a number of criteria and then provides financing for those projects. Projects are repaid through tolls, taxes, or fees.

Public–private partnerships (PPPs) are continuing to grow as an alternative financing method for projects. In a PPP, the local government enters into a contract with a private company to complete an infrastructure project. Both parties provide financing for the project, so it is not the sole responsibility of the local government to pay for the project. A project completed through a design–build–operate method allows the local government to let a project be financed through a private company. The private company is required to finance, design, construct, and operate the project, but the public is able to use the facility. The private company is able to recover its investment through the collection of revenue from the project.

It helps an engineer to understand the sources of funds available to local governments. Not only will you impress local government officials with your knowledge, but you'll also be able to speak to your officials in their terms. It is great to be able to discuss funding sources with local officials, to demonstrate your commitment to being a partner in getting their projects off the ground.

* * *

Vivian spends most of the morning, the lunch hour, and part of the afternoon thinking about the relationships among the city, the county, and the road. She keeps wondering about how her project is being paid for. She realizes she has never thought about project funding on a large scale before. In the past when she worked on the project, it moved forward, she was aware of the budget, her firm got paid, and the contractor got paid, but she did not give a second thought to where the money to pay for the project came from.

It is bugging her enough that she tracks down Tom to visit with him about it. She wants to know where the money is coming from for the project she is currently working on. Tom is happy to chat about project funding and pulls out a file with old paperwork on the job. He thumbs through the file until he finds the sheets that detail how the road project will be funded. He quickly walks to the copy machine to make a copy to give to Vivian.

When he returns they both look at the sheet. The city is using three funding sources to pay for the road. They are taking money out of their street maintenance fund, funds generated from taxes on property owners in the city limits. They are receiving money from the gas tax allocation they get from the county. They are also borrowing money from a state program set up to loan money to municipalities for infrastructure projects. Seventy-five percent of the project will be paid for by the street maintenance fund, 15% will come from the gas tax, and the final 10% will come in the form of loan funds. Vivian finds the information to be straightforward and is a little surprised it has taken her this long to ask about how the projects she works on get funded.

Learning Local Government: Who and How

Now that you have had a chance to see a broad overview of local government, it is important to dive into some of the details. Civil engineers often interact with the people and processes of local government as their careers advance, and it helps to know the elected leaders and follow local politics. When you know about the local elected officials, you can get them to pay more attention to presentations when you touch on locally relevant subjects they care about. It is also valuable to get a feel for how the local government decision-making process works and how best to feed information to the decision makers.

One of the best ways to learn about local government is to attend the meetings or watch them on local public access television during the week. They are not as stimulating as a crime drama or a sporting event, but they will provide you with a wealth of knowledge about the people and the processes involved. Watching the meetings and gathering info before presenting will help you prepare an effective presentation. Another good place to learn about local government is through websites and social media feeds. Websites often provide biographical information on local elected leaders, and you should be on the lookout for leaders who are engineers or interested in infrastructure. Social media feeds provide insight into the types of meetings that take place, how often, and who is interested in that information.

Do not forget to research influential local government staff members who interact with elected officials or are closely involved with infrastructure. These are often public works directors or city engineers. Knowing who they are, how long they have been in their positions, and how they relate to the local elected leaders is valuable. It is always helpful to know which local elected leaders are ardent supporters of their areas of town so you can discuss how a project benefits their constituents. The same holds true for a leader who is a strong advocate for certain issues, such as the environment or transportation. If you know these things, then your presentation can include information on how the project is good for the environment or improving mobility.

Each of the different forms of local government comes with a unique process that is used to debate and decide issues. Governments strive to have a uniform set of steps to process information and deliberate decisions. If you are working with a local government, it is important to understand how things progress. There is a series of steps to get an infrastructure project from the beginning of the local process through approval by the local governing body and into construction. Often the best way to learn the process is to work through it a couple of times; with experience comes a better understanding for how to navigate it in the most productive way.

Local governments that follow a good process are able to reach a good decision or the best decision possible when there are no good decisions. While this is helpful to the local government body, it can be frustrating for engineers. Few things can drive an engineer crazier than working through a long, drawn-out public process. However, it is important to understand that a good process can ensure all the bases are covered before deliberations with a local government body move too far along. If done correctly, the process should detect land mines that could blow up the project before a lot of time and money have been invested in it.

A few of the benefits of a good local government process are:

- It ensures that stakeholders (e.g., neighbors and concerned groups) are consulted and encourages discussion on the project during the design phase.
- It allows local government staff a chance to review the project and discuss project aspects that are favorable or unfavorable.

- It encourages all stakeholders to take a chance to voice their support of or opposition to the project at several different points during the process.
- It allows for the project to be changed or updated to meet concerns expressed during project discussions.

* * *

Because Vivian has been curious about the county commissioners, Tom gives her the lowdown on some of them. He pulls up the county's website to show Vivian the commissioner's page. District 1 Commissioner Mary Smith is keenly interested in traffic safety issues because her nephew was crippled when hit by a car. District 2 Commissioner Joe Jones is an accountant, and he scrutinizes all issues by revenue and cost. He is currently working with the sheriff to pursue a new jail. District 3 Commissioner Marty Mason owns a travel agency, so airport issues always get his attention, but he is also active in senior citizen's activities. Tom adds that, from his viewpoint, none of them know much about water or sewage, so you cannot rely on one of them to be an advocate for that issue. Tom's perspective is based on a water and sewer project he had to work through with the county about a year ago. Luckily, Vivian is working on a road project, so Mary and Marty will be interested.

Tom relates to Vivian some of the frustrations he encountered on that project and over the years working through the local government process. At the beginning of his career, he felt that his engineering designs were all that was needed for a successful project, and he resented anyone who was not on board with his design. He did not enjoy the process of public meetings that poked and prodded his ideas and helped shape them into something on which the stakeholders could agree. After a few years of fighting the local government process, Tom decided to embrace it as a way to build a solid foundation for his projects. Tom tells Vivian she will need to have patience with the process or it will drive her crazy. He was glad he had not expected a clean and easy process with the water and sewer project, because he got what he expected.

Case Study: The Engineer versus the Neighbors, or Subdivision Access versus Sewer Alignment

A small, late-developing subdivision was being proposed in an older part of town. A few longtime residents adjacent to the subdivision were not exactly against the development, but not exactly wild about it either.

The residents knew the area would build up eventually, but when the day actually came, they felt that they should have a lot of influence over the subdivision's development, specifically concerning the access road location.

When approval to build the subdivision came before the city council, the neighbors spoke up to voice their concern about traffic safety at the subdivision access point. The engineering study determined that the best place for access would be directly across from three of the existing homes. A few hundred feet to the north of the proposed access point was a small hump in the road that could potentially cause sight-distance problems for cars exiting the subdivision, but the engineering study determined it was not a safety hazard. The neighbors did not believe the report and felt the access should go much further south because of the hump in the road and the danger caused by it.

The neighbors quietly and quickly conceded they were not wild about the road being across from their houses, but that was not their greatest concern. This statement revealed the neighbors' true intentions (to impede and influence the development), but they quickly dropped the argument and moved back to their main point. They were at the meeting to protect the public from the bad idea this engineer proposed for the access road placement. They were advocates for the public good and wanted to do what was safe for the travelers.

The engineer spoke up to point out that the access was located in compliance with city regulations. The subdivision's water and sewer mains would be installed under the internal roads per city policy. In order for the sewer main to flow under gravity, the internal roads needed to be located as designed, and the access to the main road must be located at this point. The engineer was asked to discuss with the council the complications that would arise if the access point was moved as requested. He tried to explain that if the road were moved, the sewer-line grade would not allow gravity flow out of the subdivision and a pump station would be needed. In his explanation, he used terms like grades, drop, effluent, and hydraulic gradient, all the while referring to engineering plan drawings. He did not realize it, but he was jeopardizing months of design work, as well as potentially causing his client costly upgrades by trying to rely on technical arguments.

When the engineer is done, one of the neighbors gets back up to tell the council that he really does not care what the engineer says or what data he produces; the intersection is not safe. He tells an anecdote about a near miss at his driveway and says that should be proof the access is not safe. He says the engineer may have his facts and figures correct, but he is not looking out for the public's safety. If he was, how could anyone disagree with the neighbors? He proceeds to tell the council the engineer is only looking out for what is good for the developer, not the city. He cannot believe that council members would authorize an unsafe intersection.

After a few more questions to prompt the engineer to explain the location of the access point in a way the council can understand, the situation is not getting better. The council needs short, fairly simple answers, not the detailed technical answers he keeps using. The engineer does his best to answer the questions in a technically competent manner. However, his answers may not hold up against the neighbor's plea for a safer intersection. To get a neutral perspective, the public works director is asked to explain what happens if the road is moved. He throws the engineer a lifeline.

He agrees with the engineer that if the road location is changed a pump station will have to be added. The cost of pump station operation and maintenance will be borne by the city, and he would prefer a sewer that flows by gravity. He can understand why the neighbors want the access moved, but he cannot justify moving the access based on engineering judgment. Because the public works director was an experienced and effective public speaker, he was able to explain the information in understandable terms, not engineering jargon. He did not abandon the facts, but he communicated in direct terms instead of with complex engineering design.

The access did not get moved, a pump station was not necessary, the engineer was happy, the neighbors were a little disappointed, and, in the end, the city council made the correct decision. But with a little less technical talk and a better narrative, the engineer could have had an easier time in front of the city council and would not have needed to be saved by the public works director. The neighbors told a compelling story; the engineer did not.

Engineers as Advocates for Infrastructure

Every few years, the American Society of Civil Engineers (ASCE) releases an update of America's Infrastructure Report Card. These report cards point to one of the great challenges facing America, replacing and upgrading infrastructure at the end of its useful life. While patches can be put on infrastructure to extend its life, the overwhelming evidence suggests that a massive investment needs to be made to upgrade America's infrastructure. This isn't news to engineers—we have known about the need to invest for many years.

Engineers can play an important role in addressing this problem by becoming effective advocates for infrastructure. They understand the importance of these assets, and they should be explaining their significance. Good infrastructure leads to a better quality of life, lubricates the economy, and creates a better atmosphere for business. Engineers need to lend their voices and expertise to all levels of government to produce the best public policy for our cities, our states, our nation, and our economy. Engineers need to be able to address audiences in a fashion that adds to the conversation and clearly delineates the vital importance of good infrastructure.

Public policy and investment decisions are only as good as the input into the process. Engineers need to have a positive effect in shaping good public policy and decision making. We cannot do that if we communicate in a way that decision makers do not understand. Government officials will have a greater understanding of the importance of infrastructure and a greater respect for engineers if engineers communicate effectively. It puts the profession in a better light and builds the credibility of all engineers.

If the necessary improvements to infrastructure are going to be completed, engineers will be central to the process. They will design and oversee construction of the infrastructure that needs to be fixed. However, before resources are allocated for construction, engineers need to do a successful job communicating how allocating time and money for infrastructure improvements makes for good public policy. To accomplish this, engineers must show the public how their services touch their daily lives and how it improves the quality of life.

You Are the Expert

When you are addressing local elected officials as a consulting engineer or as an advocate for infrastructure, you should be prepared for how you will be received by local elected officials. On the whole, local elected leaders will treat engineers with esteem. While the engineering profession is not as glorified in many quarters as medicine or the law, most local elected leaders give credence to engineering training and expertise. They are not likely to challenge an engineer's technical expertise. However, that does not mean they have nothing to offer in the way of advice to an engineer. They are more likely to take engineering advice if an engineer takes their advice in return. Local officials are not going to tell an engineer what type of asphalt to specify in a road design, but they can probably provide valuable insight about where to hold the community meeting on the project and who should be in attendance. Listen to their advice on these matters because this is their area of expertise.

While local elected leaders are likely to listen to an engineer as a technical expert, you can lose that goodwill in a hurry if you do not treat them with respect. It is never good practice to speak down to the council or to give a presentation that is pretentious. While they probably do not know as much about engineering as you do, your job is not to prove you are smarter than the local elected official. Dazzling them with your high IQ is not going to win points and will get you pegged as arrogant and inflexible. You may think you are inspiring confidence, but you are probably coming off as superior. These are not the traits you want to show local elected leaders. Remember, you do not know as much about banking as a retired banker or as much about housing as a real estate agent. You will find a vast range of intelligence and expertise in local government, and if you speak to them as a "know-it-all" engineer, you will ruin your credibility as a technical expert.

You Can Be a Spokesperson

There are several opportunities for engineers to engage or help a local government. These avenues provide good chances to assist the community and work on public speaking. Most local governments have boards and commissions that provide governing functions or give advice. Check the local government website to determine the requirements to serve on a local government board. Another option is to visit with a local elected official and let the official know you are interested in serving on a commission and ask how to go about applying for the board. There are dozens of ways to get involved with local government, and they are often looking for people to fill those roles.

Serving in a local government role is a great way to practice public speaking. If you become a zoning board member, then you will be able to watch others give testimony to the board. Think about what they did well, what went wrong, and how you can use that information the next time you give a presentation. You will also be called on at meetings to give a short speech on your feelings about the issue presented to the board. This experience will give you opportunities to provide a well-communicated message to those listening to board deliberations.

Another opportunity to become involved with local government is working on a campaign. Campaigns are always in need of volunteers and will give you a chance to watch politics up close, advise the candidate on infrastructure issues, and give you an opening with the candidate. Campaigns are usually flexible about allowing you to volunteer when you can, so think about becoming involved the next time that politician you support is running for office.

However you choose to become involved with local government, you will find it to be a rewarding experience and a great way to give back to your community.

Case Study: Over the River and through the Woods to the Municipal Wastewater Treatment Plant

Many municipalities are bordered by exurban communities—unincorporated towns with a few thousand residents. They have enough housing density to provide some utility services to their residents, but are not often up to the same level of service as the adjacent city. During my career on the city council, we dealt with an unincorporated area that has a water and sewer district to provide drinking water to most of the residents. However, all wastewater systems in the area are underground treatment systems; most of them are simple septic tank and drain field treatment systems.

The water and sewer district had been trying to provide a central sewer system for a number of years, but was unsuccessful in getting the residents to vote for the money needed to develop the system. Part of the problem was a state law that required water and sewer districts to get a 60% approval vote before they could sell bonds for the system. Over the course of several years, the district was successful in getting well over 50% of the people to vote for the sewer, but they could never quite get to the 60% level. Eventually the state law was changed so that they would only need a 50% vote to sell the bonds to create a central wastewater system.

Before the successful bond passage, an engineering evaluation of the alternatives for providing sewer service showed the best alternative option was to connect to the adjacent city's existing wastewater system. This would provide the greatest economic and time-sensitive solution to the problem. The district negotiated an agreement with the city to provide wastewater treatment if the district provided the collection system. They also received a federal grant to construct the sewer main across the river. However, by the time the bond issue passed, the agreement with the adjacent city had expired and needed to be renewed. This put the grant in jeopardy of lapsing because of the failure to construct the central sewer collection system in a timely manner.

When the agreement came up for renewal, representatives from the district were unable to attend the city council meeting. With a different city council from the one that voted for the initial agreement, there were many questions asked at the meeting about the renewal agreement. Since no one was at the meeting to answer those questions, the agreement was delayed until they could be answered at a subsequent meeting. The agreement was brought up at the next meeting, and members of the district were there. However, the voice of opposition to the agreement also arrived, urging the city not to approve the agreement.

The area encompassed by the water and sewer district boundaries is home to several people who were in strong and vocal disagreement with the need for a central wastewater treatment system. They were an independent bunch who saw no need for the proposed wastewater collection system. There was also some sentiment expressed for the construction of a wastewater treatment plant by the district, with no need to use the adjacent city's wastewater treatment plant.

Many city council members also aired their concerns about the agreement. They felt little need for the city to provide wastewater treatment to the area. They were concerned that problems could arise that would make it tough for the city to take a hard line with the district. They saw little benefit to the city in treating the wastewater from the district.

The meeting answered a few questions, but the outcome of the renewal agreement remained uncertain. The next step in the process was to vote on the agreement at the next regular council meeting. After a long hearing about the agreement, the council decided on a 6 to 5 vote to enter into an agreement with the district to treat their wastewater should they develop a collection system.

Most of the discussion at the meeting centered on politics, conjecture, and risk, with little discussion of the engineering or environmental aspects of the project. The engineer did a good job of explaining the engineering aspects of the project and the council was not ill-informed about them. One minor aspect about the small effect the extra wastewater would have on the city's treatment plant was missed, but the discussion was a good reflection of how the council felt about the agreement. Little of that feeling had to do with the engineering aspects of the agreement.

This was a clear case where engineering took a back seat to politics. Based on engineering judgment, the agreement made sense for both the city and the district. However, engineering judgment was far down on the list of considerations. More important were the economic, cultural, and environmental issues tied to the agreement.

The story of the sewer expansion shows that even if an engineer does communicate well, sometimes politics trumps engineering. Politics plays a role in infrastructure, and there are times when politics outweighs the good communication skills of an engineer. However, the absence of good communication skills would have doomed this project. Given the contentious nature of the project, poor public speaking by the engineers involved could have swung the votes against the project. While good communication was not the main reason a good infrastructure decision was made, it was an invaluable aspect of the discussion.

Epilogue

Early in January, Vivian sits in Tom's office for her yearly evaluation. Tom tells Vivian he is impressed with her work overall, but the item that impressed him most was how hard she worked at improving her public speaking skills. He notes she did particularly well in speaking to the city council about the Main Street road reconstruction project by preparing for the presentation, practicing hard, and speaking well. He lets her know that her work did not go unnoticed, and Vivian is proving to be a valuable asset at the firm. She is an engineer they can trust with more responsibility for interacting and communicating with clients. If she continues to improve and impress with her communication skills, she will advance in the firm.

Ask most managers at engineering firms what sets an engineer apart, and they often cite the engineer's ability to communicate ahead of technical ability. Almost all engineers have a level of competency in engineering that is on par with their peers. Rarely will an engineer be singled out for being a better engineer than the one sitting in the office next door. Good public speaking skills serve you, the engineer, over the course of your entire career and certainly add to your attractiveness to any organization. It will help in getting new jobs and moving up the career ladder.

Engineers who can communicate successfully with nonengineers have a skill that will be noticed and sets them apart from other engineers. Engineers who can communicate effectively with clients and the public, provide leadership, work within multidisciplinary teams, understand how governments function, and build long-lasting relationships with clients are the engineers who succeed and become the leaders in their fields and firms.

Appendix: Resources to Understand Local Government

International City/County Management Association (ICMA) (http://icma.org/en/icma/home)
Founded in 1914, ICMA is the premier local government leadership and management organization. Its mission is to create excellence in local governance by advocating and developing the professional management of local government worldwide. In addition to supporting its more than 10,000 members, ICMA provides publications, data, information, technical assistance, and training and professional development to thousands of city, town, and county experts and other individuals throughout the world.

National Association of Counties (NACo) (http://www.naco.org/)
NACo is the national organization that represents county governments in the United States. Founded in 1935, NACo improves the public's understanding of county government, assists counties in finding and sharing innovative solutions through education and research, and provides value-added services to save counties and taxpayers money.

National Conference of State Legislatures (NCSL) (http://www.ncsl.org/)
The NCSL is a bipartisan organization that serves the legislators and staffs of the nation's 50 states, commonwealths, and territories. NCSL provides research, technical assistance, and opportunities for policymakers to exchange ideas on the most pressing state issues.

National Governors Association (NGA) (https://www.nga.org/cms/home.html)

This bipartisan organization of the nation's governors promotes visionary state leadership, shares best practices, and speaks with a unified voice on national policy. Founded in 1908, the National Governors Association is the collective voice of the nation's governors and one of the most respected public policy organizations in Washington, D.C. Its members are the governors of 48 states, two commonwealths, and three territories.

National League of Cities (NLC) (http://www.nlc.org/)
NLC is the oldest and largest national organization representing municipal governments throughout the United States. Its mission is to strengthen and promote cities as centers of opportunity, leadership, and governance. Forty-nine states have organizations to represent cities in their states.

U.S. Conference of Mayors (USCM) (http://usmayors.org/)
The USCM is the official nonpartisan organization of cities with populations of 30,000 or more. Each city is represented in the conference by its chief elected official, the mayor.

Index

advocating for infrastructure, 147–152; engineer as expert, 148–149; engineer as spokesperson, 149–150
anxiety, 110–114
appearance, physical, 115–116
art of public speaking, 5–6
assessing speaking skills, 94–99; feedback from others, 95–97, 96*f*; learning by observing, 97–99; self-assessment, 94–95
assessing the setting, 36–46; backup plan, 42–45, 43*f*; boardroom setup, 40*f*; classroom setup, 39*f*; evaluating the space, 38–42; finding the space, 38; physical attributes, 37–38; theater setup, 41*f*
audiences: boards and commissions, 28; characteristics, 24–25; engaging, 117–120; identifying, 21–32; interactivity, 119; interdisciplinary teams, 30–31; internal vs external, 29–30; local government officials, 27–28; non-engineers, 25–27; participation and questions, 118–119; the public, 32; reaction to speaker, 1–4; recommendation to an audience outline, 58–59, 62*f*; relationship with, 82; with shared sensibility, 31; sophisticated vs unsophisticated, 29; understanding, 23–24
audiences with shared sensibility, 31
avoiding jargon, 116–117

backup plan, 42–45, 43*f*
boardroom setup, 40*f*
boards and commissions, 28
body language, 107–110; facial expression, 109–110; gestures, 108–109; posture, 108
breathing, 106
building an outline, 55–59

career day, 130
case studies: anxiety, 113–114; complete streets, 34–36; death by PowerPoint, 87–89; delivery of speech, 113–114, 122–123, 130–132; local elected officials, 138–140; missing opportunities to speak well, 130–132; municipal wastewater treatment plant, 149–150; Murphy's water tank, 45–46; overcoming butterflies, 113–114; promotion not received, 8–9; speaking strategies, 122–123; subdivision access, 145–147; technical proficiency vs communication skill, 122–123; TEDx speech, 59–60; three-minute speeches, 69–78
characteristics of audience, 24–25
city council speech outline, 57–58
city government, 135–137; city council speech outline, 57–58; commission, 137; council with a manager, 136; mayor and council, 135–136; representative town meeting, 137
civil engineering viewpoint, 7–8
classroom setup, 39*f*
collecting material, 53–55
commission, 137
commission and administrator, 138
conference presentations, 47–48
connecting with the audience, 119–120
council and executive, 138
council with a manager, 136
county government, 137–138; commission, 137; commission and administrator, 138; council and executive, 138

death by PowerPoint, 87–89
debt financing, 141
delivery of speech, 6, 91–132; case studies, 113–114, 122–123, 130–132; design to delivery, 92–93; missing opportunities to speak well, 130–132; overcoming butterflies, 113–114; speaking habits, 123–132; speaking skills, 94–99, 100–114; speaking strategies, 114–123; technical proficiency vs communication skill, 122–123
design, 51–90; design to delivery, 92–93; outline to speech, 60–78; outlining, 53–60; planning to design, 52–53; visual aids, 78–89
design to delivery, 92–93
direct payment, 140

elected officials, local, 138–140
engaging the audience, 117–120; audience participation and questions, 118–119; connecting, 119–120; interactivity, 119; the opening, 118
engineer as expert, 148–149
engineer as spokesperson, 149–150
entertaining presentation, 19
evaluating the presentation space, 38–42
eye contact, 100–102

facial expression, 109–110
factors in speech plan, 11–12
feedback from others, 95–97, 96*f*
financing mechanisms, 140–143
finding the presentation space, 38
focused practice, 124
funding mechanisms, 140–143

gestures, 108–109
getting to know local government, 133–134
glossophobia, 5
government committee meetings, 48
government hierarchy, 134–140; case study, 138–140; city government, 135–137; county government, 137–138; special purpose district governments, 138

handouts, 79–81
hierarchy of government, 134–140; city government, 135–137; county government, 137–138; special purpose district governments, 138

identifying the audience, 21–32; audience characteristics, 24–25; audiences with shared sensibility, 31; boards and commissions, 28; interdisciplinary teams, 30–31; internal vs external, 29–30; local government officials, 27–28; non-engineers, 25–27; the public, 32; sophisticated vs unsophisticated, 29; speaker's role, 22–23; understanding the audience, 23–24
informative presentation, 17–18
interactivity, 119
interdisciplinary teams, 30–31
internal vs external audience, 29–30

learning about government, 143–144
learning by observing, 97–99
local government, 133–154; advocating for infrastructure, 147–152; case studies, 138–140, 145–147, 149–150; debt financing, 141; direct payment, 140; funding mechanisms, 140–143; getting to know, 133–134; government hierarchy, 134–140; learning about, 143–144; local elected officials, 138–140; meetings, 47; municipal wastewater treatment plant, 149–150; officials, 27–28; public-private partnerships, 142; resources, 153–154; special purpose district governments, 138; subdivision access, 145–147; Tax Increment Financing, 141–142; viewpoint, 7
lunch presentations, 127

mayor and council, 135–136
meeting types, 46–50; conference presentations, 47–48; government committee, 48; local government, 47; one topic, many meetings, 49–50; proposal meetings, 49; sales meetings, 49; staff report on safety, 48
meetings of local government, 47
missing opportunities to speak well, 130–132
municipal wastewater treatment plant, 149–150

non-engineers as audience, 25–27

officials from local government, 27–28
one topic, many meetings, 49–50
opening of speech, 118

outline to speech, 60–78. *see also* outlining; case study, 69–78; persuasive presentation, 75–78; project progress presentation, 72–75; section development, 63–67; service club presentation, 70–72; starting with a story, 67–68; structure of speech, 68–69; talking points vs word-for-word, 60–63; three-minute speeches, 69–78
outlines. *see* outline to speech; outlining
outlining, 53–60. *see also* outline to speech; building an outline, 55–59; case study, 59–60; city council speech outline, 57–58; collecting material, 53–55; recommendation to an audience outline, 58–59, 62*f*; technical conference speech outline, 56–57; TEDx speech, 59–60, 61*f*
overcoming butterflies, 113–114

participation and questions from audience, 118–119
persuasive presentation, 18–19, 75–78
physical attributes of presentation space, 37–38
planning, 11–50; assessing the setting, 36–37; backup plan, 42–45, 43*f*; case studies, 34–36, 45–46; factors in speech plan, 11–12; identifying the audience, 21–32; meeting types, 46–50; moving to design, 52–53; objective of presentation, 13–17; planning checklist, 15*f*; preparation vs procrastination, 19–21; room attributes, 37–41; scope of presentation, 12–13; selling your expertise, 32–34; type of speech, 17–19
planning checklist, 15*f*
planning to design, 52–53
posture, 108
PowerPoint. *see* presentation software
PPPs. *see* public-private partnerships
practice, 124–127
preparation, 6, 19–21
presentation, objective of, 13–17
presentation software, 81–87, 86*f*, 87*f*; death by PowerPoint, 87–89; effective slides, 85–87, 86*f*, 87*f*; relationship with audience, 82; software choices, 81; software defaults, 83–85; visual aids, 87–89
presentation space: assessing the setting, 36–46; backup plan, 42–45, 43*f*; boardroom setup, 40*f*; classroom setup, 39*f*; evaluating the space, 38–42; finding the space, 38; physical attributes, 37–38; room attributes, 37–41; theater setup, 41*f*
procrastination, 19–21
project progress presentation, 72–75
proposal meetings, 49
public as an audience, 32
public speaking: art of, 5–6; audience reaction to speaker, 1–4; case study, 8–9; civil engineering viewpoint, 7–8; delivery, 6, 91–132; design, 51–90; glossophobia, 5; local government, 133–154; local government viewpoint, 7; overview, 1–10; planning, 11–50; preparation, 6
public-private partnerships, 142

QSB. *see* qualification-based selection, speaking about
qualification-based selection, speaking about, 129

reaction to speaker from audience, 1–4
readiness, 106–107
recommendation to an audience outline, 58–59, 62*f*
relationship with audience, 82
representative town meeting, 137
resources for understanding local government, 153–154
room attributes, 37–41

safety presentations, 127–128
sales meetings, 49
scope of presentation, 12–13
section development, 63–67
self-assessment, 94–95
selling your expertise, 32–34
service club presentation, 70–72, 128
setting, assessing, 36–46; backup plan, 42–45, 43*f*; boardroom setup, 40*f*; classroom setup, 39*f*; evaluating the space, 38–42; finding the space, 38; physical attributes, 37–38; theater setup, 41*f*
situational awareness, 120–121
slide deck. *see* presentation software; slides, effective
slides, effective, 85–87, 86*f*, 87*f*
software. *see also* presentation software: choices, 81; defaults, 83–85

sophisticated vs unsophisticated audience, 29
speaker's role, 22–23
speaking classes, 128
speaking habits, 123–132; case study, 130–131; focused practice, 124; missing opportunities, 130–131; practice and experience, 126–127; practicing out loud, 124–126; where to practice, 127–130
speaking skills, 100–114; anxiety, 110–114; assessing, 94–99; body language, 107–110; case study, 113–114; eye contact, 100–102; feedback from others, 95–97, 96*f*; learning by observing, 97–99; overcoming butterflies, 113–114; self-assessment, 94–95; voice control, 102–107
speaking strategies, 114–123; avoiding jargon, 116–117; case study, 122–123; engaging the audience, 117–120; physical appearance, 115–116; situational awareness, 120–121; technical proficiency vs communication skill, 122–123
speaking when no one else will, 128–129
special purpose district governments, 138
speeches: case study, 69–78; city council speech outline, 57–58; delivery of, 91–132; entertaining presentation, 19; factors in speech plan, 11–12; informative presentation, 17–18; opening of, 118; outline to speech, 60–78; persuasive presentation, 18–19, 75–78; project progress presentation, 72–75; section development, 63–67; service club presentation, 70–72; starting with a story, 67–68; structure of, 68–69; talking points vs word-for-word, 60–63; technical conference speech outline, 56–57; TEDx speech, 59–60, 61*f*; three-minute speeches, 69–78; type of, 17–19
speed of speaking, 105–106
staff report on safety, 48
story, starting with a, 67–68
structure of speech, 68–69
subdivision access, 145–147

talking points vs word-for-word, 60–63
Tax Increment Financing, 141–142
technical conference speech outline, 56–57
technical proficiency vs communication skill, 122–123
TEDx speech, 59–60, 61*f*
theater setup, 41*f*
three-minute speeches, 69–78
TIFs. *see* Tax Increment Financing
tone and pitch of voice, 103–105
type of speech, 17–19; entertaining presentation, 19; informative presentation, 17–18; persuasive presentation, 18–19

understanding the audience, 23–24

viewpoint of local government, 7
visual aids, 78–89; case study, 87–89; handouts, 79–81; presentation software, 81–87, 86*f*, 87*f*
voice control, 102–107; breathing, 106; readiness, 106–107; speed, 105–106; tone and pitch, 103–105; volume, 105
volume of voice, 105

where to practice, 127–130; career day, 130; lunch presentations, 127; safety presentations, 127–128; service club presentation, 128; speak about qualification-based selection, 129; speak when no one else will, 128–129; speaking classes, 128

About the Author

Christopher A. "Shoots" Veis, P.E., M.ASCE, has 20 years of experience as a civil engineer focusing on municipal engineering assignments involving water and wastewater systems, land development, permitting, and project management. He developed a unique perspective during five years of service as an elected member of the city council of Billings, Montana's largest city, taking an active role in budgeting and infrastructure projects. He is an engaging, experienced public speaker delivering presentations on multiple topics to professional conferences, public meetings, and classrooms, including a TEDx speech. On Friday nights, he finds stress relief officiating high school football and acting at Billings Studio Theatre.

Veis is a registered professional engineer with a bachelor's degree in environmental engineering and a master's degree in project engineering and management from Montana Tech of the University of Montana. He was named a Top 20 Under 40 construction professional by *ENR Mountain States* magazine, a Rising Star in Civil Engineering by *CE News* magazine, and the Outstanding Young Engineer by the Billings Engineers Club. He is married with two children.

RECEIVED
FEB 16 2018